SOCIOLOGISTS
IN *Action*

*This book is dedicated with deep love and admiration
to the memory of Eleanor D. and Allen D. Willard. We are
forever grateful for the unwavering support you gave us,
the pride you expressed in us, the dreamer you nurtured in
us, the ways you challenged us to think more deeply, your
gift for filling our gatherings with humor, the stories and
heritage you shared so generously, your quiet appreciation,
your gratitude for all of the gifts that life and love bring, and
most particularly, for providing us with the most remarkable
example of true, joyful, thankful, enduring love.*

Sail on!

SOCIOLOGISTS
IN *Action*

SOCIOLOGY, SOCIAL CHANGE, AND SOCIAL JUSTICE

EDITORS

Kathleen Odell Korgen
William Paterson University

Jonathan M. White
Bentley University

Shelley K. White
Worcester State University

Los Angeles | London | New Delhi
Singapore | Washington DC

Los Angeles | London | New Delhi
Singapore | Washington DC

FOR INFORMATION:

SAGE Publications, Inc.
2455 Teller Road
Thousand Oaks, California 91320
E-mail: order@sagepub.com

SAGE Publications Ltd.
1 Oliver's Yard
55 City Road
London, EC1Y 1SP
United Kingdom

SAGE Publications India Pvt. Ltd.
B 1/I 1 Mohan Cooperative Industrial Area
Mathura Road, New Delhi 110 044
India

SAGE Publications Asia-Pacific Pte. Ltd.
3 Church Street
#10-04 Samsung Hub
Singapore 049483

Acquisitions Editor: David Repetto
Editorial Assistant: Lauren Johnson
Production Editor: Eric Garner
Copy Editor: Diane DiMura
Typesetter: Hurix Systems Pvt. Ltd.
Proofreader: Sally Jaskold
Indexer: Jeanne Busemeyer
Cover Designer: Gail Buschman
Marketing Manager: Erica DeLuca
Permissions Editor: Karen Ehrmann

Printed in the United States of America.

A catalog record of this book is available from the Library of Congress.

978-1-4522-0311-9

This book is printed on acid-free paper.

MIX
Paper from
responsible sources
FSC® C014174
www.fsc.org

13 14 15 16 17 10 9 8 7 6 5 4 3 2 1

Contents

Acknowledgments

We are enormously grateful for the unwavering and enthusiastic support of David Repetto. As well as being an exceptionally gifted acquisitions editor, he is truly a great guy with whom it is wonderful to work. We know how fortunate we are to have him as our editor. We are also very appreciative of the organizational work, constant support, and guidance of our original editorial assistant, Maggie Stanley. Eric Garner, the production editor for the 2nd edition, also deserves our thanks and praise. Diane DiMura expertly copyedited the 2nd edition and it was a pleasure to work with her. It is always enjoyable to work with the SAGE team.

Shelley would like to thank her parents and siblings for their constant, loving support. I am also especially grateful to my nieces and nephews for keeping me playful even as I pursue my research and action on social injustices in the world. I am ever indebted to my grandparents for lending me the perspective and appreciation that long lives well lived afford. I also must thank my mentors who have supported and guided me on my path to understanding the marriage of scholarship and activism, including Charlie Derber, Eve Spangler, Bill Wiist, Pauline Hamel, Kris Heggenhougen, Monica Onyango, Bill Bicknell, Lucy Honig, and Bob Woods. I am also grateful to the many inspiring young activists I have met through Free The Children, who in many ways teach me more than I could ever teach them! Finally, I feel so fortunate that my life partner is also my partner in changing the world. Thank you, Jonathan, for supporting me and journeying with me every day!

Jonathan owes a special debt of gratitude to his mentors and colleagues Charlie Derber, Eve Spangler, David Karp, Gordie Fellman, Irv Zola, Morrie Schwartz, Karen Hansen, Sue Dargan, Lucille Lawless, Joe Bandy, Terry Arendell, Craig Kielburger, Marc Kielburger, Fintan Kilbride, and his students at Bridgewater State University for their incredible guidance and support on his journey as a sociologist in action. I am especially grateful to my students and to the Free The Children and Me to We staff and youth, past and present, who have and continue to inspire me with their deep commitment to social change and social justice. I am eternally grateful to my family for their constant support and unconditional love. Especially, I could never express deeply enough how lucky I feel to have my wife Shelley by my side as my best friend and partner in life. Thank you, Shelley, for inspiring me with your passion for social justice and equity, and for this incredible journey we are on together!

Kathleen is grateful to have earned her PhD in sociology at Boston College, where she learned that sociology can and should make a positive impact on the world. She thanks her family for their love, support, and patience. Mom, you will always be my #1 editor. Thanks for all you do. Julie and Jessica, the stories in this book will help you to understand more fully why your mom loves being a sociologist. You two make me so very proud to be your mom. Jeff, thanks for being such a wonderful motivator, source of inspiration, and all around incredible partner (and excellent dad).

Finally, we are enormously indebted to our contributors—an incredible array of inspiring Sociologists in Action, and gifted writers. They make us proud to be sociologists!

Introduction

Have you ever wondered . . .

- How you can make a positive impact on the world?
- What sociologists do and what *you* can do with a sociology degree?
- How you can use sociological tools to help create social change?
- Why some of the most amazing people are sociologists?

If so, you have started reading the right book! In the following pages, you will learn how sociologists are using sociological tools in a wide variety of social justice efforts in the United States and across the globe. In each chapter, inspiring sociologists share stories of how they have used sociology to understand and influence the world around them.

As outlined in the Table of Contents, the chapters cover the key topics in sociology courses. The discussion questions at the end of every chapter will spark interesting and nuanced discussions, grounded in the "real world" work of sociologists. We also have provided great web-based resources at the end of each chapter and on the *Sociologists in Action: Sociology, Social Change, and Social Justice* website for those who want to delve further into the topics covered.

As sociologists, we are always on the lookout for patterns and there are some key patterns that one can find right here in this book! As *public* sociologists, all of the sociologists in action featured in this book use sociological tools to make a tangible impact on society. They each fulfill what Randall Collins (1998) calls the two core commitments of sociology: developing and using the sociological eye and engaging in social activism. They (1) use their "sociological eyes" to see beneath the surface of society to notice and examine patterns of injustice *and* (2) actively confront and alleviate those injustices. All, too, use what C. Wright Mills (1959) described as the "sociological imagination" to connect personal troubles to public issues. Many describe in their pieces how their own experiences with inequality made them notice patterns of inequality across society and become eager to use sociological tools to address them.

After recruiting the contributors for *Sociologists in Action*, the three editors of this book are more convinced than ever that many of the most amazing people in the world are, indeed, sociologists. We are sure that you will feel the same way after reading their stories. We also hope that you will become inspired to follow in their footsteps and become a Sociologist in Action yourself. Sociology is an exciting discipline that contains the tools for tackling many of the social issues facing us today. We know you will gain inspiration from the examples in this book of how sociology can be used for social change and social justice, and we urge you to use the sociological tools *you* are learning in your course to make a positive impact on society!

References

Collins, R. 1998. The sociological eye and its blinders. *Contemporary Sociology,* 27(1), 2–7.

Mill, C. W. 1959. *The sociological imagination.* Oxford, UK: Oxford University Press.

Chapter 1
The Sociological Perspective

In this chapter, four very different Sociologists in Action pieces provide examples of how the sociological perspective can be used to understand society and make a positive impact upon it. Each of the four stories illustrates the two core commitments of sociology, using the sociological eye to notice social patterns and utilizing social activism to address social issues. The authors also describe how they used their sociological imaginations to relate their personal experiences to larger social issues and to their work as sociologists.

Starting off this chapter with "Sociology: Promise and Potential Through Praxis," Cheryl Joseph vividly describes how her childhood experience of temporarily moving to Los Angeles and falling from a comfortable working class to a lower-class lifestyle "forced [her] to remove the blinders of familiarity and look past a way of life [she] had assumed was normal." This new perspective allowed her to start developing a sociological eye and notice patterns in society of which she had been unaware. It also gave her the drive to use her sociological imagination to critically analyze the world around her and to find ways to "put sociology into action" and address inequities in society.

In the second piece in this chapter, "Human Rights and the Sociological Imagination: How Sociologists Can Help Make the World a Better Place," Mayra Gomez's work as a human rights advocate illustrates how the tools and perspective of sociology can both illuminate and help efforts to alleviate injustice. Gomez uses her sociological background, to great effect, in her work with the Global Initiative for Economic, Social and Cultural Rights and other human rights organizations. In doing so, she has helped create "concrete policy change" and a "more socially just world."

"Stand Up and Speak Out," Judith Wittner discusses how she enabled her students to "look beyond the information given out by news sources . . . to question the motives of people in positions of authority, to learn why society operates as it does, and to *act* on it." In ceding the traditional power of the professor in the classroom, she revealed that the individual "classroom experience . . . [is] part of a politics of knowledge affecting subordinate groups more generally." She argues that such awareness "is a critical first step toward building a citizenry liberated from mainstream media's control of political ideas and actions."

In the final piece of this first chapter, Georgette Bennett shows how she has used sociological tools "as a change agent and 'action' sociologist" during the course of her life. Throughout her many successful careers, Bennett has utilized her sociological background to make innumerable important, positive impacts on our society. Her body of impressive work includes: helping to create the first sex crime unit (now popularized on "Law and Order: SVU") and changing the systematic unequal treatment of women in the NYPD; successful careers in broadcast journalism, public relations, and marketing; and founding the Tannenbaum Center for Interreligious Understanding. In every venue, Bennett has used "the unique tools of our trade to make an impact on the world."

SOCIOLOGY: PROMISE AND POTENTIAL THROUGH PRAXIS

Cheryl Joseph

Notre Dame de Namur University, Belmont, California

Sociology professor Cheryl Joseph received her doctoral degree from Wayne State University in Detroit where she was born and raised. There, Dr. Joseph began her advocacy against racism, sexism, poverty, militarism, urban demise, and environmental degradation. Her 20-year position with a major airline taught her about corporate operations in the global economy, as well as different cultures throughout the world. A faculty member of Notre Dame de Namur University (NDNU) near San Francisco since 1988, her proudest contribution is the Animals in Human Society concentration that she created within the sociology major at NDNU. Dr. Joseph recently published *Dealing With Difficult People: It's a Zoo Out There* (2010) and contributed a chapter to *Teaching the Animal: Human–Animal Studies Across the Disciplines* (2010), edited by Margaret DeMello.

Until the age of 13 (in 1961), I lived in a comfortable working-class neighborhood of Detroit, surrounded by factory-working fathers and stay-at-home mothers with their flocks of postwar babies. Included in this lifestyle was a stellar educational system and excellent health care benefits made possible by the struggles of the labor union to which my dad belonged. All this changed suddenly, however, during what Dr. Phil might call a "defining moment." My father, deadened by the stultifying effects of assembly-line work and lured by his brother's offer of a new start as a small businessman, unceremoniously quit his job and uprooted the family to begin a seemingly promising life in Los Angeles, California.

After selling the family home and driving cross-country, we arrived in Los Angeles to begin our new lives. Regrettably, however, my uncle abruptly backed out of the deal immediately upon our arrival. This left my parents and me in an environment where we knew no one, had less than one year's saving to live on, and had only my dad's skill as a television repairman to depend on. Being more of a socialist than an entrepreneur, my father had no ability to run a business. As such, our lives quickly spiraled downward. We had moved from a spacious Cape Cod home with a sizeable yard to a cramped apartment, from a neighborhood where children played safely outside after dark to one where streets were lined with porno houses and derelicts.

After a summer spent exploring this new city by bus (rather than the bike I was accustomed to) with a cousin 2 years younger and 10 years more mature than I, junior high school beckoned. I found myself, sans cousin, in a friendless setting without my familiar clique. I resolved to take the initiative and make friends. I was more lucky than adroit in this venture as there were dozens of kids just like me—new to Los Angeles because their families sought a better life and desperate for friends. I recall one girl I met that first day before classes had even begun. In those first moments, "Pamela" told me she and her family, originally from rural Arkansas, were living in their car until she could "make it" in the movie industry. I saw Pamela only sporadically after that first day and after a few months, not at all. Each time, she looked dirtier, gloomier, and more stressed. As far as I know, she never did make it to the silver screen. We were joined that first day by "Kathleen," who told us she was put on a bus in Idaho and sent to live with relatives in Los Angeles because her mother could no longer afford her. Later, I met "Sandy," whose mother and siblings depended on the income she derived from prostituting herself after school. Occasionally, "Sandy" would show me the bruises she incurred from an abusive john.

On the other hand, the same school that served this underclass also attracted students from the wealthier side of the city. I became friends with "Rachel," who would be delivered to school each day by the family's driver. Rachel would invite me to her home in the sumptuous canyons where her family "dressed" for dinner, and I learned to enjoy foods I had never heard of before. I became accustomed to hearing my classmates discussing the movie stars who were living in their neighborhoods.

At the same time, the meager savings on which my own family depended were dwindling quickly. After two years, we returned to Detroit with $60, my dog, and whatever we could fit into the back of the station wagon. We ate at truck stops instead of the family-style restaurants where we had dined on the trip west. Instead of Holiday Inns with swimming pools, we slept in places that can only generously be described as "dives."

Back in Detroit, we were forced to live with relatives in a very small house. With 10 of us and the family dog in tight confines, tensions were inevitable. My family soon separated until we could afford a place of our own. My dad stayed with his sister, my mom with a good friend, and I got farmed out to whoever would take a 15-year-old and her dog. I moved a lot that summer. I did not know at that time that my family was, by definition, homeless. Life improved, however, when my father was hired back at his previous job. With all the limitations of factory work, it nonetheless (thanks to a strong union) provided my family the benefits of home ownership and a college education for me.

I tell my story not because I enjoyed this trip down memory lane, but because all of these experiences laid the foundation for my life as a sociologist and for the kind of sociology I practice. These encounters forced me to remove the blinders of familiarity and look past a way of life I had assumed was normal. In order to understand the lives of my new-found friends, I had to critically examine their worlds and my own. I learned to appreciate and empathize rather than criticize. These life events helped me develop my sociological perspective. At the same time, I found the ability to connect personal troubles to public issues, what C. Wright Mills termed the "sociological imagination," invaluable to understanding the connections between such social structures as the economy and the individual problems that my friends, family, and I incurred.

It was these lessons, in addition to the sociology-as-action approach, that I wanted to convey to my students when I began teaching sociology. Students in my Social Problems classes, for example, spent time in soup kitchens and homeless shelters engaged in participant observation. As part of a county-wide census, one project took them to the streets of San Francisco after midnight to count those individuals sleeping in doorways and cars, and on park benches or pavement. In another class, students were required to simulate a day in the life of a mother on welfare. With only $5 and a doll that represented an infant-in-arms, they had to navigate the neighborhood using only public transportation. They were instructed to find the nearest welfare office, buy groceries for the day, and go to the elementary school where their oldest child purportedly had a discipline problem. One of my students summarized her experience thusly: "I've always believed there is no reason for child abuse. But I found myself slamming that damn doll on the ground when the bus driver would only accept exact change and I had to go from store to store to get it. What an eye-opener!" Another said, "I gave up from sheer exhaustion."

Students in my Urban Sociology classes acquaint themselves with different social classes, cultures, histories, and such social concerns as crime and affordable housing, through interactions with the people who live in San Francisco. For nearly 20 years, my students have hosted a picnic in San Francisco's Golden Gate Park on Thanksgiving Day where they share the lunch they prepared with the 350+ homeless people who live there.

In addition, I teach a two-semester internship class in which students assist such populations as jail inmates and their families, domestic abuse victims, people living with HIV/AIDS, at-risk youth, and homeless families. One student reported, "This internship lets me see the world outside of the classroom. All the theories and concepts we learn in class come alive."

By 2000, I began to recognize the close bond many people share with their companion animals and to notice similarities in how animals and oppressed human populations like people of color, women, and the poor are treated. As a result, I created a sociology major with a concentration in animal–human studies. This Sociology: Animals in Human Society major—the only one in the world to date—includes six classes taught from the perspective of engagement and action-oriented sociology. The Animal–Human Bond course, for example, requires students to conduct research by interviewing groups with different viewpoints on an animal-related issue like breed-specific bans. They then devise a means to reach a target audience. Finally, the general community is invited to an evening where students present their messages. In the past, the presentation of these messages has ranged from developing brochures that encourage owners to neuter and spay their pets to a script for a television infomercial depicting the link between animal abuse and violence against people. This strategy helps students see the role of social awareness in social change. Students also obtain "hands-on" experience at humane societies, rescues and sanctuaries, facilities that utilize therapy animals, and organizations that advocate for animal rights.

In the fall of 2009, I undertook the codirectorship of our Dorothy Stang Center at my university. I feel honored to be part of an effort that carries forth our namesake's quest for social justice, community engagement, and environmental sustainability. For over 20 years, Sister Dorothy Stang worked with peasant farmers of the Brazilian rainforest, defending their land rights against the rapacity of cattle barons and the logging industry. Just as she began to make progress against that country's power elite, she was viciously murdered by a hired assassin. In her memory, the Dorothy Stang Center embarks on projects that create social awareness and social change both locally and globally. On the local level, we have initiated a community garden on campus and we give the produce to a nearby homeless shelter. On a larger scale, a group of us attended a vigil/protest to advocate the closure of the School of the Americas/Western Hemisphere Institute for Security Cooperation (SOA/WHINSEC), which trains the militaries of dictators in Central and South America who then massacre their own citizens.[1] In 2010, we spent Spring Break in an impoverished region of Jamaica. There, we engaged with Jamaicans to create empowerment groups that will eventually produce collectives for artists, micro-financing opportunities, and chances to build healthy communities.

My early experiences and the context that sociology provided for understanding them allowed action-oriented sociology to evolve readily for me. Marx's theories of class consciousness and social conflict resonated with me as I thought of the experiences of my friends and family. Through learning how to use my sociological imagination, I recognized the pervasive influence of social institutions on individual lives. The feminist approach of Jane Addams awoke me to the strength of women in community. The connections she created between the Chicago School of Sociology and Hull House (the settlement house in Chicago she cofounded) inspire me to combine practice with theory. Putting sociology into action fuels my passion. It is this passion that I try to convey to my students.

HUMAN RIGHTS AND THE SOCIOLOGICAL IMAGINATION: HOW SOCIOLOGISTS CAN HELP MAKE THE WORLD A BETTER PLACE

Mayra Gómez

Global Initiative for Economic, Social and Cultural Rights

Mayra Gómez is Co-Executive Director of the Global Initiative for Economic, Social and Cultural Rights. She received her PhD in sociology from the University of Minnesota, where her work was supported by a grant from the

---

[1]For more information about the School of the Americas, see James Hodge and Linda Cooper, Disturbing the Peace: The Story of Father Roy Bourgeois and the Movement to Close the School of the Americas (Maryknoll, NY: Orbis Books, 2004); Chris White, "Roy Bourgeois' mission to close SOA," *National Catholic Reporter*, 41 (February 11, 2005).

MacArthur Interdisciplinary Program on Global Change, Sustainability and Justice. She has worked with leading international human rights organizations and UN human rights bodies, particularly in the area of women's human rights, and travels widely to all regions of the world. She has authored over thirty human rights articles, books and reports, and is a regular contributor to the *Netherlands Quarterly of Human Rights*. Mayra has also served on the Amnesty International USA Board of Directors.

My love of sociology and of the social sciences came early to me as a student. In high school, I loved the ways in which sociologists thought about the most compelling and difficult issues of our time and helped to illuminate the human condition in all of its many facets. From poverty to war, from inequality to exclusion—even from fashion to entertainment—no subject was too big or too little for sociological reflection and inquiry.

Around the same time, I also developed a love of international issues, human rights, and social justice work. As a 16-year-old, my favorite movie was *Gandhi*, while everyone else's seemed to be *Top Gun*. For me, Tom Cruise's swagger paled in comparison to Gandhi's principled and stalwart opposition to the British Empire. I became a student member of Amnesty International (even now, over 20 years later, I am still closely involved with Amnesty), wrote urgent action letters on behalf of political prisoners after school, and dreamed at the time that I would go to law school to become an international human rights lawyer. I believed deeply, and still do, that social change requires social activism, and that this loop begins with the agency of people working together to make the world a better place.

Concern over human rights is, to be sure, shared by a lot of people and it has altered the international political landscape. We hear about human rights almost daily—about the right to freedom of expression, or the right to nondiscrimination, or the right to health care. It has transformed, in a fundamental way, how we think about the integrity and dignity of the human person.

In the end, I chose not to go to law school, although I was fortunate enough to take several human rights law classes through the University of Minnesota Law School. Instead, I decided to get my PhD in sociology because I really wanted to learn more not only about the legal dimensions of human rights abuse, but about the social and human ones as well. I believe that the perspectives which sociology gave me have been extremely valuable to my intellectual and moral development as a human rights activist.

So, you may be asking, what exactly have those perspectives been? Well, there are several and, quite honestly, I don't think it's possible to understand the world today, much less how to change it, without a sociological imagination. First, sociology has helped me to understand the concept and context of human rights abuse in a unique way. Sociology helps illuminate the social patterns of inequality, conflict and exclusion that underlie all human rights violations, as well as the way rights themselves are constructed and understood.

For example, sociology gives me a good foundation to understand the social systems of inequality on which human rights abuses are themselves based. The Rwandan genocide as it unfolded would not have been possible but for the social construction of ethnic identity. The crimes against women that we see throughout the world—from violence within the home, to denial of education, to unequal rights to housing, land and property—would not be possible if not premised on an ideology of gender which separates women and men on the basis of their social roles and elevates the status of the latter. All of the human rights abuses which we see in the world are by definition human-made, and that makes them ripe for sociological analysis. As Mills wrote, the sociological imagination is ". . . the vivid awareness of the relationship between experience and the wider society."[2] Even what seems like the most intimate of human experiences, from being tortured in a dank and dark jail cell, to not being able to go to school because you are a girl, are all embedded in a complex web of social process, power and norms. To change the way societies operate at a fundamental level—which is what we try to do as human rights advocates—means that you need to understand the deeper sociological forces at play and how they work.

Sociology also helps us to understand the notion of "rights" themselves. While the very idea of human rights is sometimes built upon the principle that these rights are intrinsic and essential, sociologists know that the moral framework of human rights is not timeless. Human rights—their language, their construction, and their meaning—are always in the process of social transformation. Sociology can help us to understand how and why rights claims change over time.

It's easy to see change by looking at U.S. history, and how rights claims have progressed over time in this nation. The Founding Fathers of this country believed that people were "endowed by their Creator with certain unalienable Rights, that among these are Life, Liberty and the pursuit of Happiness," but of course the caveats were that you needed to be a white, property-owning male to have such rights. Later, we saw the struggles against slavery, and then for women's suffrage, civil rights, women's equality, and the rights of indigenous people and LGBT people. This link is critical because rights and their protections are only expanded through social struggle, through people coming together and standing up for the rights of all human beings.

Sociology has also taught me that societies are complex, even counterintuitive and self-contradictory. Societies are not static, and it is not a good

---

[2]C. Wright Mills, *The Sociological Imagination* (New York, NY: Oxford University Press, 1959).

idea to generalize too much, lest we fall back on misleading stereotypes and exaggerated caricatures. As a sociologist, I find people are really fast and loose with their generalizations: "Women are like this, men are like that," or "Africans are like this, Asians are like that." I am very careful, I find, about what I am comfortable saying is really a part of "human nature." I think what defines us as human is our keen ability for learning (perhaps our most important adaptation). Our sociological eyes also help us to understand that it is through our interactions with others that our rights can either be realized or, tragically, denied.

Lastly, sociology has also given me the practical skills to engage in the day-to-day work of human rights advocacy. From a methodological standpoint, a great deal of human rights work is geared towards research, both quantitative and qualitative, and here the skills of the sociologist are very apt. Rigor in research is always needed to formulate the right questions, to know how to interpret your research findings, to understand how your research could be biased and to take measures against that to ensure its independence. All of these things you learn as a sociologist and apply in practice as a human rights advocate.

These tools and perspectives, taken together, have helped me to change the world in some very tangible ways. For example, while at the Centre on Housing Rights and Evictions (COHRE), an international human rights organization based in Geneva, Switzerland, and advocating for the right of everyone, everywhere, to adequate housing, we prepared a report on women's housing rights within the context of domestic violence. The report highlighted the plight of countless women who are unable to leave their abusers because they have no place to go. As one woman from Argentina told us: "I did not even have the money to pay rent and no money to meet my expenses. These are factors that exacerbated my circumstances of life. He (my abuser) would ask me, where was I going to go?" In the report, we argued that governments must respond effectively and coherently to the pervasive problem of domestic violence, and they must ensure that women fleeing violence have an adequate, safe housing alternative. My use of sociological tools and my sociological imagination helped to inform this report and, hopefully, to fortify the rights of those women.

The report, dealing with Argentina, Brazil, and Colombia, was very well-received and was covered by BBC Mundo[3] and BBC Brasil, and also other news outlets, including Diário do Grande ABC, Gazeta Digital, MaisComunidade.com, O Tempo, O Progresso, Abril, and Pagina 12

---

[3]See for example: http://www.bbc.co.uk/mundo/america_latina/2010/07/100716_violencia_domestica_estudio_rg.shtml

in Argentina. It was also picked up by the Web sites of Women's UN Report Network (WUNRN), Habitat for Humanity, Amnesty International Canada, The Network for East-West Women, and Ajudaparamujeres.com.

Ultimately, as result of our effort, the Municipality of Rosario (the third largest city in Argentina) amended their Program on Housing Construction for Low Income Families. That city is now setting aside 10% of the housing created through the program for victims of domestic violence, and currently there are plans to take similar steps at the federal level. It was powerful to see some concrete policy change as a result of our report and to see how my sociology background is helping to create a more socially just world![4]

In my work with Amnesty International USA, I served on their Women and Human Rights Steering Committee for four years. While I was there, Amnesty launched its "Stop Violence Against Women" campaign. We saw several successes during the time of the campaign, including here in the United States with the Senate and House Committees on Appropriations recommending increased funding for the Violence Against Women Act (VAWA). In Mongolia, their Parliament unanimously passed a law against domestic violence. Even in small countries, like the Solomon Islands, we saw success with the country's first sexual assault unit created and funding for programs relating to violence against women prioritized.[5]

All of this shows how the sociological perspective goes hand in hand with human rights advocacy. By their professions, sociologists must become adept at thinking relationally—between, for instance, the national and the international, or between the national and the local, or between the international and the local. So, too, must human rights advocates. As articulated by the renowned French sociologist Pierre Bourdieu, sociology must develop a

> scientific humanism which refuses to split existence into two realms, one devoted to the rigors of science, the other to the passions of politics, and which labors to put the weapons of reason at the service of the convictions of generosity.[6]

For me, that is why I have loved sociology, and that is what being a sociologist is all about!

---

[4]OAWEB, 'Viviendas para mujeres víctimas de violencia,' 20 December 2010. See: http://www. otrosambitosweb.com.ar/despachos.asp?cod_des=26853

[5]Amnesty International Australia, Six Years that Changed the World, 28 June 2010. See: http:// www.amnesty.org.au/svaw/comments/23283

[6]Pierre Bourdieu and Loïc J.D. Wacquant, An Invitation to Reflexive Sociology (Chicago, IL: The University of Chicago Press, 1992), 86.

STAND UP AND SPEAK OUT

Judith Wittner

Loyola University, Chicago, Illinois

Judith Wittner is a professor in the Department of Sociology, Loyola University, and is affiliated with and a former director of the Women's Studies/Gender Studies Program. She specializes in the sociology of gender and teaches ethnographic methods. She earned a PhD from Northwestern University and also pursued graduate studies in anthropology and political science. In 2006, she was awarded the Feminist Mentoring Award from Sociologists for Women in Society (SWS). Dr. Wittner has published research on domestic violence, homelessness, foster care, and factory work and has conducted workshops in ethnographic methods at universities in Lithuania, Nigeria, and El Salvador. The second edition of her book, *Gendered Worlds* (with Judy Aulette), was published by Oxford University Press in 2012.

"That's a myth!" I snapped, when during a heated class discussion of war in the news, one student claimed that civilians "spat on" returning Vietnam-era soldiers. "The only spitting I heard about was the time that pro-war supporters spat at Jane Fonda, who was against the war." It was mid-semester in the afternoon section of the course, Mass Media and Popular Culture, in the spring of 2006, and I was losing control over the class. The course was one that I had taught many times before. In previous classes, I attempted—as responsible teachers are expected to do—to keep my political thoughts to myself. By 2006, silence seemed impossible and counterproductive. How could I have no opinion about U.S. complicity in what many human rights groups have since called war crimes (Pincus, 2006)? I was eager to engage students in discussions about the part that mainstream media played in condoning the abuse and torture of Iraqis and supporting war and militarism.

No distanced and professorial approach seemed possible or, for that matter, ethical. I wanted to talk with my students about these atrocities. I wanted to make them feel the anger I felt when our elected leaders engaged in and justified such practices. Most of all, I wanted to teach them to defy the authority of these leaders and reject the media messages supporting them. This stance was my version of public sociology. As faculty members, students are our publics, but they can be a passive public, silent and willing to go along in the classroom where they put their real lives and their ideas on hold. I believed that if my students knew what Washington was doing in their name, they would overcome their passivity and challenge authority. They would "stand up and speak out" (Brecher & Smith, 2006). I hoped my students would be able to fulfill the two core commitments of sociology: developing a sociological eye enabling them to see beyond the media and to question motives of those creating our collective knowledge, and engaging in social activism to use their sociological eyes to create change (Collins, 1998).

To encourage participation, I had promised a grade of "A" to anyone willing to speak up in class. Some students took me up on my offer, and the classroom became the site of many debates. Throughout the class, I worried whether it was wrong to trade grades for participation. This day, I worried that my perceptible anger during the spitting debate undermined the spirit of this bargain. Before we met again, I did some research and found that there had been no contemporary reference to incidents of spitting, suggesting that the story was apocryphal (Lembke, 1990). But I also found an active debate on the web over this issue, with countless testimonials about the verbal abuse of returning soldiers by antiwar activists.

The semester's experiment showed that despite my wish to create a democratic classroom, I had strong emotions invested in being the powerful professor who enlightens her misinformed students. It had certainly felt normal and natural to exercise the authority that came with my position. Despite my ambivalence, however, I had opened up a space for students and they were beginning to occupy it, although not in ways I had anticipated. In fact, the first dissenters seemed, from my vantage point, to disrupt our class by their refusal to see things my way. They weren't the rebellious, progressive young men and women I had imagined would be part of a participatory classroom. Quite the opposite: The conservative students had been the first to find their voices. What was new for them was that they were speaking their minds about convictions they held silently in other classes. One of these students commented,

A lot of teachers who even though they have such a liberal or such a rightist biased viewpoint, you don't have huge class discussions. You know you're there, you're taking notes, you're listening to their lecture. The things that you agree with you underline, the things you don't agree with you strike through.

Another student was surprised by these exchanges, "because this is the first time I've had conservatives in class. I mean I know conservatives outside the class but I've never seen them speak out in class." Students in this class moved from detachment and indifference regarding classroom discussions to active participation in them.

Similarly, some male students "felt attacked" when I showed two provocative videos: *Killing Us Softly 3,* a film examining disturbing gender stereotypes in advertising, and *Tough Guise,* about the social construction of dominant masculinities and men's violence in the media. Their objections surprised another student who said, "We watched that in my Sociology 101 class and *none of the guys said they didn't like that*" (emphasis added). In this class, however, they spoke their thoughts to and with others in the classroom, and the ensuing debates moved many on both sides of the arguments to shift their positions.

The talk in class also spilled out of the classroom. A student reported,

> After class, we ended up talking about not that it was brought up in class but we were wondering what it would be like to talk about it in class, the Supreme Court case that was up yesterday.[7]. . . So we really ended up talking about that for the rest of the walk.

Such changes are hard to measure, but they also mark small beginnings of active citizenship.

Given my political perspective and those of the usually vocal students, it isn't surprising that many of the dissenting voices came from the right. The most significant impact of the class was that students challenged my authority directly. One day I made a point, now forgotten, that led a student to explode with "That's ridiculous!" I complimented him on his disrespect for authority. One student wrote on the course evaluation, "A teacher that remains neutral wouldn't be able to stimulate discussion nearly as effectively." Another "loved the fact that I could state my opinion and do so freely. . . . I think that it is important for a student's education and growth." Another wrote, "I've learned to question authority."

The fact that classrooms are hierarchies in which students are subordinates is not news to you. Slightly less obvious is the politics of knowledge that is part of that hierarchy. Disciplinary perspectives and the view from above have legitimacy, while the knowledge held by those less powerful, like students, often seems to have no place in the curriculum. Understanding the classroom experience as part of a politics of knowledge affecting subordinate groups more generally provides a sociological tool in confronting official knowledge and in searching for the subordinated and unofficial news that we as citizens need to know. We can then use the sociological understandings we have gained to work for social change and justice.

---

[7]Voting 8–0, the Supreme Court upheld a federal law that says colleges and universities must grant military recruiters the same access to students as other potential employers or else lose federal funding.

When I gave up control over the classroom, my students in turn gave up the stolid silence that protected them from engagement with others on important social and sociological questions. They stood up and spoke out. It spilled over from the classroom into their daily lives. I am convinced that such dialogue, whether I agree with the sentiments expressed or not, is a critical first step toward building a citizenry liberated from mainstream media's control of political ideas and actions. In the classroom, my students began to develop a sociological eye and to engage in social activism. They learned to look beyond the information given out by news sources with political agendas, to question the motives of people in positions of authority, to learn why society operates as it does, and to act on it.

References

Brecher, J., & Smith, B. (2006). Where are the good Americans? *The Nation.* Retrieved from http://www.thenation.com/article/where-are-good-americans

Collins, R. (1998). The sociological eye and its blinders. *Contemporary Sociology, 27*(1), 2–7.

Lembke, J. (1990). *The spitting image: Myth, memory, and the legacy of Vietnam.* New York: New York University Press.

Mills, C. W. (1959). *The sociological imagination.* New York, NY: Oxford University Press.

Pincus, W. (2006, October 5). Waterboarding historically controversial. *The Washington Post.* Retrieved from http://www.washingtonpost.com/wp-dyn/content/article/2006/10/04/AR2006100402005.html

Strauss, A. L. (1959). *Mirrors and masks: The search for identity.* Piscataway, NJ: Transaction.

GETTING BEHIND THE HEADLINES AND GOING WHERE THE ACTION IS: MY CAREER AS A SOCIOLOGIST IN NON-ACADEMIC SETTINGS

Georgette F. Bennett

Tanenbaum Center for Interreligious Understanding

Georgette Bennett, PhD, is the founder and president of the Tanenbaum Center for Interreligious Understanding. She is also a long-time consultant to the Milstein real estate, banking, and philanthropic organizations, where she develops social responsibility programs. She managed a $1.3 billion budget for the New York City Office of Management and Budget and had a successful banking career, while earning a graduate business degree. A former faculty member of the City University of New York, Dr. Bennett also taught in New York University's School of Education. In addition to a distinguished broadcasting career, she is the author of 4 books and over 50 articles in scholarly, professional and popular publications. Among many honors, Dr. Bennett has received awards from the American Society for Public Administration, International Council of Christians and Jews, Center for Christian-Jewish Understanding, New York City Comptroller, Friends of the United Nations, and Women of Wealth and was nominated for the Templeton Prize.

"Vassar Sociologist is a Lady of Action." That was the headline of a story in the *New York Daily News* on September 3, 1967. The reporter wrote:

> We spotted Georgette recently in the midst of an angry crowd in the Kew Gardens Criminal Court Building. She had read in the newspapers that supporters of 17 indicted members of the Revolutionary Action Movement, a militant Negro civil rights group, were expected to attend a pretrial hearing for the 17 . . . About 200 supporters showed and so did Georgette. "I was aware of the possibilities of danger," she explained later, "but you can't get a realistic point of view . . . from textbooks . . . alone. I'm interested in seeing things first hand and unearthing my own truth."

And that, at age 20, was the start of my life as a change agent and "action" sociologist. My journey took me through a range of careers. In all of them, the tools of sociology have given me the opportunity to create meaningful and long-lasting change.

At Vassar, I majored in sociology, drawn by its focus on inequality, social change, and competing realities. Vassar seniors had to write a thesis as a requirement for graduation. Newly armed with Max Weber's theories of social change and bureaucracy, I decided to base my research on an observational study of Timothy Leary's famous commune, which resulted in my thesis: "LSD: The Institutionalization of a Social Movement." There, despite the group's espousal of spontaneous joys, such as "turning on with the cows," I found a developing social structure with clearly defined roles and procedures.

I continued with my sociology studies in graduate school at NYU. My preferred research method remained participant observation. I burrowed behind the headlines again—this time, school decentralization and the searing black-Jewish conflict spawned by the Ocean Hill-Brownsville experiment in giving control of schools to local communities at the expense of the centralized Board of Education. This led to my dissertation, "Educated Fools or Uneducated Schools: The Social Validation of Reality in a New York City Ghetto School."

A year prior to completing my dissertation, I embarked on the standard tenure-track teaching path for which almost all doctoral programs prepare their students. During my second year of teaching at the City University of New York (CUNY), I was invited to join the Women's Advocacy Committee (WAC), an elite group of a dozen activists: Eleanor Holmes Norton, Betty Friedan, Eleanor Guggenheimer, Ronnie Eldridge, Carol Greitzer, and other icons of the women's movement. I was only a few years removed from my first year at Vassar, where I had been required to read Freidan's revolutionary book, *The Feminine Mystique*, for first year orientation. In my world, these women were legends.

WAC organized the first women's march in New York and subsequently, John Lindsay, then the mayor of New York City, designated it an official mayoral committee, mandated to do advocacy work on behalf of women in city agencies. My colleague, Ellen Mintz, and I were assigned to work with one of the last bastions of manhood: the New York City Police Department (NYPD). Our mission was to address the plight of women as victims, criminals, and colleagues in the NYPD.

One of our first official acts was to meet with police "brass." We brought with us Susan Brownmiller, author of *Against Our Will: Men, Women and Rape*. Out of that meeting came the first sex crimes unit, now popularized as *Law and Order: SVU*. Many decades later, when I was on the set of *SVU*, I got a kick revealing to the show's stars, Mariska Hargitay and Christopher Meloni, that they were looking at one of the early advocates for the unit that gave them their long-term TV franchise.

Initially, the bulk of our work focused on female police officers, who in the 1970s were confined to matron duty and secretarial work. There were only a couple hundred women in a force of 35,000—and such assignments made it nearly impossible for them to progress up the ranks into positions of authority. As women, academics, civilians, and change agents, Ellen and I had four strikes against us. In order to beef up our street "creds," we spent hundreds of hours on patrol with police officers, who resented our presence, in some of the most active precincts in New York. Once again, I was doing participant-observation research.

In order to create systemic change in the treatment of female employees in the NYPD, we had to take on the masculine, man-of-action ethic that permeated police culture. We lobbied to change the civil service requirements for policing: height requirements were abolished, civil service lists were merged so that men and women were hired from the same lists, physical tests were revamped to include only those items that were bona fide occupational qualifications. Our efforts had a major impact. Today there are 6200 female, sworn officers in NYPD—with many in command positions—comprising 18% of the service.[8]

Our ultimate goal was to edge NYPD from a law enforcement orientation to a service orientation. We did so by tackling the police socialization process, implementing a total-systems approach that included the entire career cycle, from day of recruitment to day of retirement: training, performance evaluation, incentives, early warning systems, and cultural awareness. We were on to something revolutionary. Then–Police Commissioner Pat Murphy (the legendary corruption-fighter in the days of the Knapp Commission and Serpico) asked us to take a leave of absence from CUNY and work as full-time consultants for the NYPD. The result was the Full Service Neighborhood Team Policing Model (FSNTPM), which became the subject of a national program funded by the federal government and a blueprint for community policing. Today, police departments all over the world use community policing.

As a newly minted sociology PhD, and freshly emerging from my first stint in policing, I was introduced to Dr. Morton Bard, a CUNY psychology professor, who had done groundbreaking work training police in family crisis intervention. Now, with a team of senior scholars, he was embarking on a new project that took on the criminal justice system in a broader way. It was on that team that I had another breakthrough idea: creating a multiservice crime victim service program. Dr. Robert Reiff, one of the scholars on the team, ran with it and created the first federally funded crime victim services center in the United States. And with that, the victim rights movement was born.

Being one of the few women in authority at NYPD, I had a great deal of visibility. This led to extensive work with the Department of Justice and police departments all over the country. But, after 18 months with NYPD, my leave of absence was over and I had to return to academe. I experienced powerful culture shock and it became clear to me that the ivory tower wasn't where I belonged. I needed to stay firmly anchored in the real world.

My work with policing hadn't been just about criminal justice administration. It produced a set of transferable skills and experiences in organizational development and management consulting. With the election of a new mayor, Pat Murphy's successor left NYPD to head up Chemical Bank's (today, JPMorgan Chase's) Operations Division. When he asked me to come with him, I took a second leave of absence and applied sociological skills to working on new models of work, such as flextime, and a range of personnel administration functions.

---

[8]"NYPD Celebrates Women in Policing," April 15, 2009. Retrieved from http://nycppf.org/html/nypd/html/pr/pr_2009_ph07.shtml

After a second—and last—return to academe, I was still seized with the need to get behind the headlines. Journalism and sociology dovetail in many ways. I wrote to Earl Ubell, news director at WNBC-TV, offering my services as an on-air criminologist—a first in broadcasting. Generally, one doesn't get a foot in the door without an agent. But Ubell, a scholar in his own right, saw the value of this kind of expertise at a time when crime was the number one preoccupation of the American public. I was assigned to develop stories and within a year, I, myself, was on the air as a reporter at WNBC in New York. A year after that, I became a network correspondent for NBC News. I brought my sociologist's toolkit: data analysis; an understanding of social forces as they impact specific social problems; and the knowledge of how reality is socially constructed. These enabled me to explode many popular myths in my reporting.

My career as a broadcast journalist took me all over the dial, including my appointment as East Coast host for a PBS series created by avuncular newsman Walter Cronkite. At the same time, I started working with the Center for Investigative Journalism, which developed stories for CBS' *60 Minutes*. I also developed stories for ABC's *20/20* and PBS' *McNeil Lehrer NewsHour*. During these years, I wrote many print pieces for a range of news outlets. The most gratifying of these were the investigative crime pieces I wrote for *New York Magazine*, with Nick Pileggi (*Wiseguy, Goodfellows, Casino*) as my mentor and editor.

Like many journalists, I eventually moved into public relations and marketing. *Public relations* is, at its core, the application of journalistic skills to creating a focused perception of a person or an issue. PR people call it controlling the message; sociologists call it managing social cues. Either way, public relations requires understanding one's market and one's client. Marketing is just another form of sociological trend analysis for the purpose of achieving a clearly defined set of results.

In 1992, my husband, Rabbi Marc Tanenbaum, died. Marc was an inspiring figure—a pioneer in the field of interreligious reconciliation and a world-renowned human rights activist. When he died, I decided to devote the bulk of my time to building on his work. I founded the Tanenbaum Center for Interreligious Understanding, which took me behind a different set of headlines—this time, the world of religion-based conflict and prejudice. All of my grounding in sociology came together: sociology of religion; survey research; networking; the socialization process; social construction of reality; trend analysis; studies of the link between religion and violence, social movements, sub-cultures, authoritarian personalities, "true believers" and more.

Borrowing from my NYPD playbook, the Tanenbaum Center focuses on the institutions that mold behavior. We also work with peacemakers who intervene, on the ground, in conflict zones around the world. Back channel work—in places like Jordan, Iraq, Syria, Oman, Israel, Nigeria—keeps me where the action is and behind the headlines, where I always want to be—a sociologist in action, using the unique tools of our trade to make an impact on the world.

DISCUSSION QUESTIONS

1. When you were reading Cheryl Joseph's piece, did you think of her as a homeless person during the transitions of her difficult early years, or did you assume she was not homeless because she always had a place to stay? How would you define homelessness, and why do you think most homeless people are without a home?

2. When Cheryl Joseph and her family moved from Detroit to Los Angeles, they also, unfortunately, moved from the working to the lower class. In speaking of the new friends she met, she states, "In order to understand the lives of my new-found friends, I had to critically examine their worlds and my own." How do you think the tools of sociology can help you to examine and better understand your world and the worlds of others? Use your sociological imagination to relate a personal trouble that you are going through to a social issue.

3. According to Mayra Gómez, what is sociology "all about"? How has her work reflected her description of the discipline?

4. Describe how the report on women's housing rights and domestic violence that Gómez and her colleagues created led to tangible, positive differences in the lives of women. Why do you think it made such an impact and how did her sociological skills help to create this impact?

5. Have you ever been in a classroom like the one Judith Wittner created where students can feel free to express their views, no matter their political bias? How do you think this type of classroom setting would enhance your understanding of society? Would you feel comfortable challenging the opinions of (a) your fellow students and (b) your professor? Why or why not? How do your experiences with power influence your answers?

6. Do you agree with Judith Wittner's statement that sometimes "[n]o distanced and professorial approach seemed possible or, for that matter, ethical" for a professor? What are advantages and disadvantages of professors not including their own viewpoints in the classroom?

7. Georgette Bennett has used her sociological training in many careers over her life. Give at least two examples from her piece that exemplify that sociological tools

can be used in a broad range of professions. Describe how you might use your sociological training in your own career.

8. Describe two ways in which Georgette Bennett and her colleagues helped transform the NYPD. How do they exemplify Bennett's desire to be a Sociologist in Action?

9. Discuss how Cheryl Joseph and Mayra Gómez provide powerful examples of what C. Wright Mills called "the promise of sociology," to connect personal troubles with public issues.

10. Which, if any, of the pieces in this chapter gave you a new perspective on the field of sociology? Why? Had you been aware of the "action" part of sociology before? Why or why not? How does that component of sociology influence your interest in the field?

RESOURCES

The following Web sites will help you to further explore the topics discussed in this chapter:

American Sociological Association (ASA)	http://www.asanet.org
Citings and Sightings	http://thesocietypages.org/citings/
Glossary of the Social Sciences	http://www.faculty.rsu.edu/~felwell /glossary/Index.htm
Sociological Tour Through Cyberspace	http://www.trinity.edu/~mkearl/
Sociosite	http://www.sociosite.net
Virtual Library of Sociology	http://socserv.mcmaster.ca/w3virtsoclib/

To find more resources on the topics covered in this chapter, please go to the Sociologists in Action Web site at **www.sagepub.com/korgensia2e.**

Chapter 2

Theory

A s you saw in Chapter 1, part of the sociological perspective entails notic-
ing social patterns in society. Once we observe those patterns, we then
need a way to make sense of them. Social theories help us to understand why
those patterns exist and why society operates in the ways that it does. Theories
also help us to figure out how to influence society and achieve practical goals.
The Sociologists in Action in this chapter each use sociological theory to help
inform their research and to work toward important social change.

In the first piece in this chapter, "Critical Mixed-Race Studies: The
Intersections of Identity and Social Justice," Andrew Jolivette advocates
using a "critical mixed-race theoretical perspective to effectively participate
in and organize for social justice across our differences." After learning
that he has AIDS, Jolivette gradually came to embrace all aspects of his
identity (gay, multiracial, male, etc.) and realize that "we must not be afraid
to embrace the multiple identities and spaces that make us who we are."
Through the critical mixed-race theoretical framework, Jolivette has been
able to connect his own experience as a multiracial gay man with AIDS
to the societal need to overcome differences and demarginalize oppressed
identities. He continues to help lead a "battle with the demons of internalized oppression that have us believing we have to change who we are to be
full citizens of this nation." Putting theory into action, he encourages us "to
continue to move beyond and across the lines that dictate who and what we
can become" and to work together for justice for all Americans.

Through "Doing Sociology: Creating Equal Employment Opportunities,"
Menah Pratt-Clarke shows how she has used both her law and sociological
training to improve hiring practices at her university. She describes how she
has used her expertise in critical race theory "to help the university to change
its policies; increase the number of positions that were posted; and enhance

its compliance with equal employment opportunity laws." The result is a fairer hiring practice which ensures that members of traditionally marginalized groups have an equal opportunity when positions become available.

In the final piece of this chapter, "Using Sociology for College Success," Laura Nichols utilizes Pierre Bourdieu's theoretical work on human, social, and cultural capital to make sense of why first-generation college students have lower graduation rates than other college students. She notes that, as Bourdieu pointed out, in order to be successful in society, one must have not only "individual abilities and skill sets (human capital)," but also "access to resources in networks of family, friends, and acquaintances (social capital)." In addition, one needs "knowledge of the parts of culture that are valued by the elite in that society such as art and literature, and experience and ease when in venues such as fine restaurants, the opera, and charity events (cultural capital)." Using this theoretical perspective, Nichols has helped encourage the creation of programs designed to assist first-generation students to attain the types of capital they need to succeed in college and beyond.

CRITICAL MIXED-RACE STUDIES: THE INTERSECTIONS OF IDENTITY AND SOCIAL JUSTICE

Andrew Jolivette

San Francisco State University, San Francisco, California

Andrew Jolivette is associate professor and chair of American Indian Studies at San Francisco State University where he is also an affiliated faculty member in Race and Resistance Studies and Educational Leadership. Jolivette is the author of three books: *Obama and the Biracial Factor: The Battle for a New American Majority* (2012); *Louisiana Creoles: Cultural Recovery and Mixed-Race Native American Identity* (2007); and *Cultural Representation in Native America* (2006). He is currently working on a new book, *Indian Blood: Mixed-Race Gay Men, Transgender People, and HIV*. Professor Jolivette currently serves as the board president of the Institute for Democratic Education and Culture—Speak Out, as co-chair of the GLBT Historical Society Board, and as board vice-chair of the DataCenter for Research Justice in Oakland, California. He is an IHART (Indigenous HIV/AIDS Research Training) Fellow at the Indigenous Wellness Research Institute at the University of Washington in Seattle, a legal expert with the Round Table Group, and an advisory board member of the 2-Spirit Grant Program at the Native American Health Center in San Francisco.

I didn't always aspire to be a college professor. However, I always had a sense of the importance of social justice because of the many conversations I had with my parents about the treatment of people of color throughout history. As a high-school student, I always thought that becoming a lawyer would be the best way to improve people's lives. It wasn't until my junior year in college at the University of San Francisco (a small, private Jesuit college with a mission of social justice) that I found a new approach to trying to transform society in my own small way.

At the time I was an English literature major with plans to apply to law school. Something wonderful changed my life, though. I took an Introduction to Sociology class and soon learned that as a sociologist I would be able to study the law while also studying society as a whole. I quickly changed my major to sociology, and after teaching Native American and women's studies at the Summerbridge Program for fifth and sixth graders at San Francisco Day School, I made the decision to apply to graduate programs in sociology instead of law school. While other disciplines tend to focus on one specific aspect of society—the mind, the body, history, gender, or race, for example—sociology offered me an opportunity to study all of these things and to try to understand how different aspects of our identities intersect to make us complete people.

As a mixed-race gay man, understanding the intersections of identity has always been important to me. The theoretical perspective of critical mixed-race studies offers a way of thinking about the world through the eyes of others. Acknowledging multiple aspects of society and utilizing multiple worldviews, this perspective develops the potential to create new frameworks that go beyond colorblind or post-racial movements that suggest we should be a nation where race, gender, and sexuality do not matter. By utilizing a critical mixed-race sociological framework, I believe we can link common struggles for solidarity together because multiethnic people often have to know more about all sides of themselves than people from one ethnic community.

How can we use a critical mixed-race theoretical perspective to effectively participate in and organize for social justice across our differences? At a recent keynote talk I gave in the Seattle area, a group of students from a lesbian, gay, bi, transgendered/gay–straight alliance (LGBT/GSA) organization asked during the question-and-answer session how they could organize in a hostile environment. I responded by saying that it is important to make connections on a personal level, where we can see each other for the full sum of who we are. I proceeded to suggest to them that they invite the Christian or college Republican student organization(s) to one of their LGBT meetings so they could discuss issues they might have in common. Perhaps they could cosponsor an event on something like health care or education that impacts both groups; and in the process of naming and claiming the issues that impact each of the groups, they might be able to find that common ground to hold multiple realities and experiences as valid. I also said that, at the very least, even if the Christian or Republican group declined the invitation, they will have made the effort to reach out across the different spaces that confine us and keep us from truly transforming society.

As a sociologist concerned with reforming, not just explaining, society I believe the greatest thing we can do is to bring greater visibility to issues that are not talked about. We must not be afraid to embrace the multiple identities and spaces that make us who we are. There is no one way to be indigenous, African, Latino, Asian, Arab, Jewish, or white. We as Indigenous Nations, as people of color, or as white antiracists have to think about what political issues we privilege and the others that we ignore. We also have to face the fact that when we privilege certain agendas in our communities, we further marginalize those most at risk—we forget about issues that create double and triple forms of marginalization for queer people of color, for veterans of color, for women of color, or for people of color living with AIDS.

I try to use myself as an example of what I mean by a critical mixed-race sociological framework. At the age of 27, I, too, once saw the world with a somewhat singular lens. I was quite focused on organizing in communities of color and on issues of racial and ethnic justice. Despite being a gay man of color, I didn't always attempt back then to see how the other aspects of my identity intersected. In the fall of 2002, I was a doctoral candidate at UC Santa Cruz, a middle school dean at Presidio Hill School, and a lecturer in sociology and ethnic studies at San Francisco State University. At the time, I was on top of the world. Sociology had opened my eyes to many fields that I could work within to improve society: I could be a teacher-activist, I could be a community organizer-advocate, I could be the change I wanted to see. I wasn't ready yet. One morning when walking up the hill from a local park with some of my students, I had trouble breathing. Another teacher said I might have walking pneumonia and should get it checked out. I thought the breathing along with a nagging cough that had lasted for several months would just go away with more time. I was wrong.

I eventually went to the emergency room, with the urging of my parents, to get checked out. Once I arrived, I felt much better and thought, "I should have just stayed home. I feel fine. They're going to think I'm just wasting their time." So as the nurse took my temperature, I fully expected to be going home. I was wrong. My temperature was 102 or 103 and my oxygen level was 87. She informed me that I was a very sick man. As I lay in a bed waiting for X-ray results, a doctor suddenly emerged and said she had some questions for me. She proceeded to tell me that I had pneumonia and it's on the front and back of my lungs, something they usually only see in the elderly and in people with HIV. My heart dropped into my stomach. She then rattled off several questions: "When was the last time you had an

HIV test? Can we give you an HIV/AIDS test?" The room seemed to get very small; it also seemed to be moving ever so slowly until it was spinning.

I stayed overnight in the hospital, and the next morning, a doctor came into my room and asked me to turn the television off. He then told me, "We have the results of your test. I'm afraid it's positive." I stopped breathing, my ears began to swell, my stomach knotted up as he talked, and it was like I couldn't hear him anymore. I started thinking about my parents, about my brothers and my sister, and about my students at Presidio Hill School. "Do you have any questions?" he asked. He proceeded to explain to me that the T-cells fight infection and that the viral load measures the amount of the virus in your blood. I was having an internal conversation with myself as I listened to him. Once I snapped out of it, I asked him what my count was. "You have 35 T-cells." "What's normal?" I asked. "500–1,800," he responded. "Oh my God, what does that mean?" I asked him. He explained that anytime you have less than 200 T-cells, you become vulnerable to opportunistic infections, and that the CDC (Centers for Disease Control) defines AIDS as anyone who has or has ever had less than 200 T-cells. "So I have AIDS?" I asked. "Yes," He answered. "What's my viral load?" He responded that my viral load was over 500,000 copies per milliliter of blood. The closer one gets to one million and the lower the T-cell count, the closer one is to death.

After being released from the hospital a few days later and many pounds lighter, I went to stay with my parents. It was tough once I got out of the hospital not knowing what to do or what would happen. Would I finish my PhD? Would I ever work again? How long did I have? And how did this happen to me? Disclosing my HIV/AIDS status to my friends and siblings was very hard, perhaps the hardest thing I've ever had to do. Every time I'd see my two younger brothers and my older sister talking and laughing and including me, I started to think, one day I might not be here to hear their laughter, to share in the jokes, to see their children.

I thought, if I'm going to die, I'm going to finish what I started first, so I worked with a sense of urgency. I sent my dissertation in to my committee that January and graduated in March of 2003. Slowly, I started putting my life back together. I'm happy to say that my viral load is currently undetectable and my CD4 count (T-cell count) is currently 700 (within the normal range). I also am in a privileged place because, unlike many people living with AIDS, I have health care, so I can pay $15 instead of $1,500 for one bottle of medication.

HIV and AIDS are things that I have learned to live with. They are also something that a part of me feels happened for a reason. I wasn't sure if I should disclose my status in this essay, but then I thought to myself, as a gay man of color, I have a responsibility to disclose. This is a very personal decision, but in communities of color, we lack faces to make this pandemic real. If you've never known someone living with AIDS, now you do. You know my story, and in sharing it, I hope that others will know that they can live with this.

Over the last seven years, I have learned that AIDS is not me; *I* am me. AIDS is only one part of my life. I share this story with my People of Color and AIDS class, and I share it with you to say that I, like all of us, have choices, and with those choices comes responsibility. I owe it to my communities to be speaking out about not just race, not just sexuality, not just education and health, but about women's rights, about immigrant rights, about sovereignty rights. I—we—owe it to our ancestors to continue to struggle for the human rights and dignity of everyone. We owe it to them to battle with the demons of internalized oppression that have us believing we have to change who we are in order to be full citizens of this nation. We must continue to move beyond and across the lines that dictate who and what we can become.

I have learned over the last seven years that if we don't stand up for one another, if we don't show up to be counted, then we will continue the cycle of exclusion and oppression. We must continue to resist. We must continue to stand up and reach out with our hands, with our bodies, with our spirits and our souls. As a sociologist, these are words that I try to live by every day of my life. I share my story all across the country to honor what I discovered so long ago in that Introduction to Sociology class—change requires not just theory, but practice. I am confident that sociology students today will tell their own stories as counselors, journalists, policy makers, professors, as so many different things, and ultimately as sociologists in action transforming the world.

For more information about Andrew Jolivette's work, please visit the Web site www.speakoutnow.org/ or www.nativewiki.org/Andrew_Jolivette.

DOING SOCIOLOGY: CREATING EQUAL EMPLOYMENT OPPORTUNITIES

Menah Pratt-Clarke

University of Illinois at Urbana-Champaign

Menah Pratt-Clarke is the associate chancellor and director of the Office of Equal Opportunity and Access at the University of Illinois at Urbana-Champaign. She received her MA in literary studies from the University of Iowa and her law degree and PhD in sociology from Vanderbilt University. An associate professor at the Institute of Government and Public Affairs, Lecturer in Law at the College of Law, and an adjunct professor in the Department of African-American Studies, she teaches in the areas of critical race, critical race feminism, and Black feminism. She recently published *Critical Race, Feminism, and Education: A Social Justice Model* (Palgrave Macmillan, 2010).

I am a Black woman. I am also a feminist and critical race scholar. I look at life through a lens that is informed by my race, gender, and class status, and I am sensitive to issues of oppression and privilege that affect all people. My life's work involves developing partnerships and collaborations to help create a fair, equitable, and just society. I do this work as a scholar-activist—using my scholarship to be an activist. The combination of my law degree and PhD in sociology helps me to use multiple ideas, theories, and strategies to solve social problems involving employment, diversity, and equity at a university.

I work in the Office of Equal Opportunity and Access at the University of Illinois at Urbana-Champaign. Our office's mission is to provide leadership in promoting and fostering an inclusive environment on campus. We oversee the implementation of the campus affirmative action plan; enforce equal employment opportunity laws; and provide workshops and programming on diversity and inclusiveness. Our office addresses issues involving constitutionally and legally protected identities, such as race, gender, disability, religion, and national origin. I spend a significant amount of time working on equal employment opportunity matters. It is our office's responsibility to ensure that every individual, regardless of identity, has an equal opportunity to apply and be considered for a job.

On a daily basis, I think about how to create change in the world using sociological and legal tools. One of the sociological tools is called the *Transdisciplinary Applied Social Justice* (TASJ) model. TASJ is "the application of concepts, theories, and methodologies from multiple academic disciplines to social problems with the goal of addressing injustice in society and improving the experiences of marginalized individuals and groups" (Pratt-Clarke, 2010, p. 27). The model draws from law, political science, sociology, history, cultural studies, critical race feminism, and critical race theory. It also uses the ground-breaking work of Patricia Hill Collins (2009). By focusing on the interconnected relationships among people, power, systems, and social structures, this model follows the belief that individuals must be at the center of any study of social problems and that their narratives and stories form the foundation for understanding the effects of social problems. From their experiences, we learn how they are affected by their own power and the power of those around them.

Power can be seen as "something that groups possess" but also as an "intangible entity that circulates within a particular matrix of domination and to which individuals stand in varying relationships" (Collins, 2009, p. 292). Collins defines four domains of power. The *hegemonic domain* involves systems of thoughts, ideologies, perspectives, values, beliefs, and stereotypes which legitimate and justify power. The *disciplinary domain* is the invisible and visible policies, practices, procedures, preferences, rules, laws, and methods of operation which often embody the values of the hegemonic domain. This disciplinary domain manages power. The *structural domain* is where the disciplinary domain and hegemonic domain operate through systems, institutions, and structures. This structural domain organizes power. Finally, the *interpersonal domain* involves the day-to-day interaction between individuals within structural domains. It is through the interpersonal domain that the hegemonic and disciplinary domains are implemented and maintained. These domains are important for engaging in praxis. Paulo Freire defines *praxis* in *Pedagogy of the Oppressed* (1986, p. 36) as "reflection and action upon the world in order to transform it." Praxis involves applying knowledge through action. It is action that ultimately creates change and that is why the TASJ model is such an important tool for change.

I used the TASJ model to review the university's hiring policies. The university's policies allowed a significant number of temporary teaching positions to be filled without a search process. This meant that many vacant positions were not posted, thus eliminating the opportunity for qualified candidates to apply. The issue that needed to be reviewed was whether there was a strong justification for waiving the search process. This is particularly important since the primary purpose of equal employment opportunity law is to create an "equal employment opportunity" for all qualified candidates to apply and be considered for positions. If open and vacant positions are not posted, then employment opportunities are not created for a broad applicant pool. I combined the TASJ model with the "Straight A" approach: *ask, analyze,* and *act*, as I reviewed the university's hiring policies, ascertained how they needed to be adjusted, and put those findings into action.

The first A step was to *ask* many questions and I used the domains as a guide. In the interpersonal domain, I found that many people were involved in the hiring process, including the hiring manager, an affirmative action officer, an equal employment opportunity officer, and search committee members. In the structural domain, I learned that there were many units and offices involved, including academic and administrative departments, colleges, human resource offices, and budget offices. I also asked questions about the disciplinary domains and the policies and procedures related to hiring. I found that there were some written university policies and state laws, and that there were also unwritten practices and procedures that guided the hiring process. Finally, I needed to learn about the hegemonic domain and the culture of the university around its hiring practices. I learned that teaching needs are very important and that the university often needed to hire faculty quickly to be able to provide sufficient instructors for courses with large student enrollments. Thus, in many instances, there was not enough time to post the position and conduct a search, as a search can sometimes take up to two or even three months. The expectation of the campus was that departments needed the ability and authority to act quickly outside of the normal process for certain positions.

The second A step requires us to *analyze* the information gathered through the first step, *ask*. There is always a great deal of information which needs to be analyzed after the ask stage. In the analysis stage, data is being reviewed and solutions are being devised and assessed. On our campus, the solutions proposed needed to balance the obligations and responsibilities under equal employment opportunity laws related to fairness and equity with the urgent campus hiring needs. In analyzing the University's policies, practices, and procedures, our office determined that the university policy needed to be revised to provide clearer guidance about when a position should be posted and when a position could be filled without a search process. It was determined that a small number of narrowly defined teaching positions could be filled without being posted, but only if there was a strong written justification and prior review and approval from our office. At the same time, it was also determined that there were other teaching positions that could and should be posted. We also clarified the length of the posting (one week or two weeks), and the scope of the advertisement (local, regional, or national).

The final A step is *act*. Social change does not happen without action and praxis. Action requires an activist—an individual who can coordinate, track issues, and bring people and teams together. At the University of Illinois, we created a committee to develop the new policy on the academic search process to clarify which positions were required to be posted. We also offered training programs and workshops to make sure that all the main actors were aware of the policy change and the reason for the change. Finally, we designed a monitoring strategy to ensure that the new policy was being followed.

Utilizing my sociology background and the TASJ model, I was able to help the university to change its policies, increase the number of positions that were posted, and enhance its compliance with equal employment opportunity laws. As a result, searches for new faculty members are open to a much broader spectrum of potential applicants, including those from minority groups historically less likely to gain positions. Through asking, analyzing, and acting, change can be made. TASJ and the Straight A system help you to be a strategic-thinker, a problem-solver, and a leader. Reviewing existing policies and creating new ones is just one example of how sociology, TASJ, and the Straight A system work together to promote social change and create equal employment opportunities. Being a sociologist in action means making a difference in the world as an agent and instrument of change, ultimately improving the lives, life chances, and experiences of others.

References

Collins, P. H. (2009). *Black feminist thought: Knowledge, consciousness, and the politics of empowerment.* New York, NY: Routledge.
Freire, P. (1986). *Pedagogy of the oppressed.* New York, NY: Continuum.
Pratt-Clarke, M. (2010). *Critical race, feminism, and education: A social justice model.* New York, NY: Palgrave Macmillan.

USING SOCIOLOGY FOR COLLEGE SUCCESS

Laura Nichols

Santa Clara University, Santa Clara, California

Laura Nichols is an associate professor of sociology at Santa Clara University in Northern California. Besides her work on first-generation college student success, her other current applied research projects focus on evaluating programs and policies that reduce homelessness and a project on institutional responses to help employees achieve work–life balance.

When I started teaching full-time, I began to notice differences among students in their patterns of behavior, classroom discussions, and conversational styles with professors, grades, and learning. These differences appeared to be based on a demographic characteristic I had never considered: whether or not someone is a first-generation college student. It turns out that your college generational status can predict a lot about your likelihood of success in college.

A first-generation college (FGC) student is typically defined as someone whose parents never attended a four-year college. (If you have at least one parent with a college degree, you are considered a continuing-generation college [CGC] student.) A national study of FGC students found that only about 25% had graduated from a four-year college within eight years of leaving high school, compared to almost 70% of CGC students (Chen, 2005). My students inspired me to study the factors that explain this disparity, and I have been conducting qualitative and quantitative studies that follow the experiences of students from their first days in college to graduation. I have also been involved in evaluating intervention programs that our college has introduced for FGC students. In my current study, my research team and I are following the experiences of FGC and CGC students who enter college with the intention of becoming medical doctors. Through surveys and follow-up interviews, we are able to better understand why FGC students have a more difficult time staying on the premed track compared to CGC students.

The different rates of success between FGC and CGC students are frequently attributed to income differences that often require FGC students to spend more time working and helping their families than most CGC students (Pascarella, Pierson, Wolniak, & Terenzini, 2004). But there are other influences as well, and social theorist Pierre Bourdieu can help us think through these various dimensions of inequality. Bourdieu (1984) thought that success in society was not just based on individual abilities and skill sets (human capital). It was also determined by people's access to resources in networks of family, friends, and acquaintances (social capital), as well as knowledge of the parts of culture that are valued by the elite in that society such as art and literature, and experience and ease when in venues such as fine restaurants, the opera, and charity events (cultural capital). Bourdieu believed that the elite value and reward these types of capital to protect their upper-class statuses.

The research I have been doing has revealed that colleges need to address all three of these types of capital—human, social, and cultural—for students to have equal opportunities for success. It also shows that FGC students must often work harder than CGC students to obtain, or at least understand, these often hidden but expected types of capital. Although structures in society mean that we are exposed to different levels of valued capital by social class, the good news is that we can increase our amounts of these types of capital in ways that can help us be successful at school and work without giving up our own cultural values. Colleges can also change those policies and practices that tend to advantage those with high amounts of capital. My research has been used to inform students about what they can do at the individual level to improve what Max Weber (1978) called their "life chances" (one's quality of life), to educate academic advisors and campus leaders, and to inform programs on campus.

Because of the segregation of neighborhoods and schools by social class and race, FGC students are more likely than CGC students to come from underperforming schools. For example, at the university where I work, with selective admissions policies, our FGC students have the same average high-school grade point averages and SAT scores as CGC students, but they have fewer opportunities to take AP and college prep classes. This often leaves them less prepared for college-level work and starting college with different levels of *human capital* than their classmates. To help with these human capital differences among our students, our university has used the results of research on FGC success to develop a bridge program that allows FGC students to start two of their classes and learn about campus resources the summer before college. The program helps students adjust to the fast pace of college as well as understand particular areas where they may need extra help. Our evaluations have found that besides helping students with academics, the program has been successful in aiding the development of other forms of capital. For example, the program helps build students' confidence and skills to take on-campus leadership positions and to talk individually with professors, thus also building their social capital.

If your family is connected to people who can help you land internships and jobs, give you advice about which courses you should take, and how to prepare for graduate school, and so forth, you have *social capital*. In our study of students who want to be medical doctors, some of our CGC students have family members who are doctors and who take them on hospital rounds, help them gain relevant experience, and even counsel them on how to do well in their chemistry classes. But schools can also play a role in helping students gain social capital. Here is an example from one of the CGC students in our study: "My [high] school holds these talks where speakers come. And during that fall, a doctor came and I really liked what he was presenting. So then I just personally talked to him . . . so he's been my mentor and if I have any questions I just ask him." Compare this to an FGC student who did not have such a program in high school but who also wants to be a doctor and understands the value of social capital: "My boyfriend just went to his dermatologist and I was like, 'Ask him how long he went to school and what he did.'" A school program combined with the first student's initiative resulted in durable social capital that he can likely use over the long term, while the second student received one-time information from a distant resource.

Colleges can help students build their career-specific social capital by connecting them to alumni and providing advising and mentoring. At the individual level, if your family and friends cannot provide the connections you need, you can make sure to find and utilize college resources and build your college networks for the future. Moreover, as a college student, you can be a form of social capital, too: Start a mentoring program with your friends and work with high-school students who want to go to college!

Finally, if you are a CGC student you are likely to have a relatively high amount of *cultural capital,* and in raising you, your parents probably practiced what Annette Lareau (2003) calls "concerted cultivation," meaning that, without realizing it, you might have been taught how to negotiate with adults, how to speak appropriately and comfortably with professionals (like faculty), and how to contest rules/policies (such as grades) if you don't think they are fair. Use of such skills is often expected in the college context, and it is also the type of training that can help you approach potential mentors and further build your social and human capital. Here are quotes from two students in our premed success study describing their interactions with professors.

> Student 1: I remember I e-mailed my [chemistry prof] once because he was wrong. . . . It's an extra credit problem and it's impossible to solve and I'm sure about that. He didn't believe me at first but now he does. I said, "That extra credit problem does not make sense. I've tried it, it's not physically possible to solve." . . . And then he wrote back to me: "No it is possible." So I tried again, and then in class I just went to him and he said, "No." So I showed him and he's like, "Oh, it's a typo."

> Student 2: I'm kind of scared of teachers. I feel it's easier for me to ask another student who understands the material than to go talk to the teacher.

Student 1 used his concerted cultivation skills to correct his professor and potentially boost his grade in the process, while the second student would have likely let the issue slide. This difference in skills and level of comfort with authority figures will likely differentially impact their success in college and in the workplace.

Although such efforts are often invisible, colleges increase the cultural capital of their students by offering structured opportunities for students to meet with faculty; through required courses in the curriculum, such as classes in art and philosophy; and by training advisors to help students navigate college. Some colleges even offer classes in golfing and wine tasting! However, administrators and faculty also need to be challenged to think about how colleges support some students better than others. For example, many college officials often presume that all students can pay tuition up front, and norms and values at our most prestigious universities follow the assumption that students should live on campus and emphasize individual over family well-being. Realizing this, our university purposely involves parents and families of FGC students in many activities throughout the year, helping them to feel comfortable on campus and encouraging them to be active parts of their children's college experiences.

I encourage you to be a sociologist in action on your own campus: Use your sociological imagination to understand your own levels of human, social, and cultural capital and, if you want, actively do things to increase yours and help your friends do the same. If the resources do not exist at your school, approach people at your college with your desire to change the way things are, and help create more opportunities for all students to succeed.

References

Bourdieu, P. (1984). *Distinction: A social critique of the judgment of taste.* Cambridge, MA: Harvard University Press.

Chen, X. (2005). *First-generation students in postsecondary education: A look at their college transcripts* (NCES 2005–171). U.S. Department of Education, National Center for Education Statistics. Washington, DC: U.S. Government Printing Office.

Lareau, A. (2003). *Unequal childhoods: Class, race, and family Life.* Berkeley: University of California Press.

Pascarella, E. T., Pierson, C. T., Wolniak, G. C., & Terenzini, P. T. (2004). First-generation college students: Additional evidence on college experiences and outcomes. *Journal of Higher Education, 75*(3), 249–284.

Weber, M. (1978). *Economy and society.* Berkeley: University of California Press.

DISCUSSION QUESTIONS

1. According to Jolivette, why should we utilize a critical mixed-race sociological theoretical perspective? Think about a social movement with which you are familiar. How might a critical mixed-race sociological perspective be useful for the members of this movement?

2. How did Jolivette's learning he had AIDS lead him to embrace a critical mixed-race framework? Think of your own life experiences. Is there anything in your own self-identity that makes you realize the need to view the world across racial/ethnic/sexual identity/class/other lines? Do you think most people have multiple identities/perspectives from which they view the world? Why or why not?

3. Describe how Menah Pratt-Clarke's discussion of the "Straight A" approach relates to the two core commitments of sociology: to use your sociological eye to notice social patterns and then to use those findings to make a positive impact on society (social activism).

4. Pratt-Clarke uses Patricia Collin's description of four domains of power in her work to advocate for fair hiring practices. How did this perspective on power help Pratt-Clarke figure out what questions to ask as she did her research and how to enact her plan of action once she developed it?

5. According to Laura Nichols, what types of capital does one need to succeed in college? Which do you think is most important for success in an institution of higher education? Why?

6. Think about your own levels of human, social, and cultural capital. Of which do you have the most? Why? What can you do to gain more of the types of capital you might be lacking? How might you go about doing so? In what ways might these types of capital be important for students (including yourself) working to create social change on campus or in the larger community?

7. Why do sociologists create theories? Using information from at least two of the three Sociologists in Action pieces in this chapter, describe how social theory can be useful for those interested in improving society.

8. Which of the theoretical perspectives/concepts described in this chapter do you find most useful in your own life? Why? Describe how you might utilize it.

RESOURCES

The following Web sites will help you to further explore the topics discussed in this chapter:

ASA (Theory Section)	http://www.asatheory.org/
Classical Sociological Theory	http://ssr1.uchicago.edu//PRELIMS/theory.html
Great Social Theorists	http://www.faculty.rsu.edu/~felwell/Theorists/Four/index.html
Sociology in Cyberspace	http://www.trinity.edu/~mkearl/theory.html
Sociology Professor	http://www.sociologyprofessor.com/
Sociosite	http://sociosite.net/topics/sociologists.php

To find more resources on the topics covered in this chapter, please go to the Sociologists in Action Web site at **www.sagepub.com/korgensia2e.**

Chapter 3

Research Methods

Sociological research methods provide us with the ability to find the information we need to create, test, and revise our theories of how society works. In order to make society better, we must first have a firm understanding of how and why it functions in the ways it does. Following the basic steps of scientific research helps us to see and measure patterns in society so that we can better understand *how* it operates. Once we have done so, we can then begin to make sense of *why* it operates that way (through critical, sociological analysis and theories). The sociologists in this chapter exemplify how good research methods can help efforts to understand and to improve society.

We can collect either qualitative (word-based) or quantitative (numerical) data, depending on our method of research. Qualitative data can come from such research instruments as interviews, participant observations, and focus groups. Quantitative data can be derived from a variety of other methods, including surveys, content analysis, and use of existing data (such as the U.S. Census). Whether we use quantitative or qualitative methods, all sociologists must follow the basic steps of the scientific process:

1. Choose a research topic.

2. Find out what other researchers have discovered about that topic.

3. Choose a methodology (how you will collect your data).

4. Collect and analyze your data.

5. Relate your findings to those of other researchers.

6. Present findings for public review and critique.

Good methods of data collection also enable us to evaluate social programs in order to determine if they are effective. In the first piece in this chapter, "The Michigan Alcohol and Other Drugs School Survey," Thomas Van Valey describes how he led an effort that provided data necessary to help design and evaluate drug education programs in the state of Michigan. As he puts it, "Carefully designed and executed survey research can tell us much about what people do and why they do it." The results, important to consider in an effort to stop drug use among children and teenagers, indicate that the War on Drugs in Michigan was not particularly effective.

Through his piece, "Using Sociological Skills for Movements to Confront Power," Bruce Nissen relates how he and a colleague established the Research Institute on Social and Economic Policy (RISEP), an institute that examines issues of particular interest to working-class and lower-income Americans. By providing "facts, figures, and information to labor unions, community organizing groups, and the media about things that impact the lives of working people: the economy, wages and benefits, housing cost, health care, immigration, and the environment," the institute has helped a wide variety of social movements advocating for social justice and worked to "transform power relationships to the advantage of our society's 'have-nots.'"

Samuel Friedman concludes the chapter with "Positive Deviance Research as a Way to Help People." In this piece, Friedman describes how he used qualitative research methods to try to determine how the members of one local union shop, "unlike the vast majority of other workers, in the Teamsters and throughout the rest of the world, were able to build an organization that could fairly and effectively defend their interests." The results, published in Friedman's book *Teamster Rank and File* (1992), have helped worker activists across the nation learn from the success of this one group of workers and use that knowledge to strengthen their own efforts for worker justice.

THE MICHIGAN ALCOHOL AND OTHER DRUGS SCHOOL SURVEY

Thomas L. Van Valey

Western Michigan University, Kalamazoo

Tom Van Valey received his MA from the University of Washington and his PhD from the University of North Carolina at Chapel Hill (1971). Having previously served on the faculties of Colorado State University, the University of Massachusetts, and the University of Virginia, he is currently a professor emeritus of sociology and former chair of the department at Western Michigan University. He has more than 50 publications to his credit, and more than 100 presentations at regional, national, and international meetings and workshops.

The Department of Sociology at Western Michigan University has a long history of applied social research, including a research center and graduate degrees with specializations in applied sociology. In 1989, President George H. W. Bush established the Office of National Drug Control Policy, and Congress passed the Drug-Free Schools and Communities Act—as part of the federal War on Drugs. The Michigan Department of Education, in response, asked our research center to develop a survey system, at relatively low cost, that would allow school districts to collect data regarding the drug use and attitudes of their students. We patterned our project after the Monitoring the Future project at the University of Michigan, which has carried out surveys of national samples of students and young adults for many years. As the director of our research center and later the department chair, I was the liaison to the state and became responsible for directing the project.

The project was essentially designed to do two things: (1) It administered drug surveys to students in Michigan public schools (as well as some private schools and schools in other states), and (2) it provided those school districts with an aggregated report of the responses provided by their students. Initially, it was intended to help school districts design their programs for drug education, and as time passed, it also became a mechanism for evaluating those programs. It was typically paid for with federal Drug-Free Schools funds, which were first administered by the state Department of Education, and later by the Michigan Office of Drug Control Policy.

The project began in the fall of 1989 and operated continuously until 2007. During its 18 years of operation, the project (known as MAOD [Michigan Alcohol and Other Drugs School Survey]) provided services to public school districts in Michigan (plus a substantial number of private schools). In addition, because the project generated a total of more than 800,000 surveys, the data became the source of several theses and dissertations, a number of publications in professional journals, and more than 30 presentations at professional meetings. It also provided a measure of financial support for literally dozens of graduate students, and part-time jobs for dozens more people in Michigan communities who helped administer the surveys.

The design of the project was simple. If a school district wanted to use federal funds to pay for information on the attitudes and behaviors of their eighth-, tenth-, and twelfth-grade students regarding substance use, they would contact the Michigan Department of Education (later the Michigan Office of Drug Control Policy) who would tell them about our project. We would establish with the school district the times and places to do the survey, and on the appointed day(s), send our graduate students or trained people from nearby communities to administer it, collect the responses, and return them to Western Michigan University for processing and subsequent preparation of the report. The instrument was professionally printed on forms that could be electronically scanned, and school personnel had no contact with the surveys at all. We even provided the pencils the students used to complete the eight-page, multiple-choice instrument.

In the beginning, the survey covered a number of familiar substances (e.g., tobacco, alcohol, marijuana, various "hard" drugs like heroin and cocaine, and various prescription drugs like amphetamines and barbiturates). We asked the students if they had ever tried the substances, how much they had used the substances in the past year, and how much they had used them in the past 30 days. In addition, we asked questions dealing with their attitudes and those of their friends and family members toward substance use, along with requesting some limited background information. After 2002, when the federal legislation was amended to the Safe and Drug-Free Schools and Communities Act, we were asked by the Michigan Office of Drug Control Policy to include questions regarding issues of safety and violence in our survey. Since these were clearly issues of considerable public concern, we added a section on safety that included questions about carrying weapons to school, property stolen or vandalized, being bullied, and being threatened or injured with a weapon. We also included some of the substances that had been introduced into mainstream culture since the survey was originally designed (e.g., methamphetamine and club drugs).

There are several reasons why the project worked so well and for so long. First, we surveyed entire populations of eighth-, tenth-, and twelfth-grade students in a district, not just a sample of them. (The only students not surveyed were those who were absent and those whose parents did not want them to participate. Typically, this was between 3 and 5% of the total student population.) This meant that the school district—and the parents—could have confidence that the results were indeed representative.

Second, the students' responses were anonymous. During the administration of the survey, their teachers did not handle the survey instruments and were not able to see their responses. No personally identifying information was asked, and we reminded the students several times that their responses were anonymous and not to write any identifying information on the survey forms. This kind of reassurance encouraged the students to be candid in their responses to the survey. In addition, only aggregated information was provided to the school district, further protecting the individual students' responses.

Third, we put the surveys through an extensive data-cleaning procedure to screen out responses that were impossible, illogical, or wildly out of range (e.g., one answer saying the student smoked marijuana once or twice in the last 30 days and another that the same student did not smoke marijuana at all in the past year). This meant that the school districts had accurate data to use in their planning.

We set up the project to be administered every two years. This way, the school districts could make comparisons of the behaviors and attitudes of nearly complete cohorts of students (i.e., the eighth graders who became tenth graders and the tenth graders who became twelfth graders two years later). Of course, not all school districts followed the procedure. Some asked us to do the survey every few years, simply to see if their situation had changed and if they needed to alter their programs to take into account potentially changing patterns of drug use. However, a number of districts did follow the procedure and were able to use the resulting data to construct drug education and prevention programs customized for the problems that were specific to their schools, evaluate the effectiveness of the programs over time, and fine-tune their design and implementation from year to year. It was particularly gratifying to see them use the results of our project, year after year, in the operation of their programs. For those school districts (and not just them), the project really did make a difference to their communities.

There were also a few districts that were interested in comparing the responses of the students to parallel information from their parents. The goal was to develop communitywide programs that would reinforce the programs taking place at the schools. Therefore, we constructed a version of the instrument that could be administered to parents and designed a sampling procedure. In each of these instances, the data analysis and report were completely customized for the particular district.

We ultimately closed down the project because of cost. Although we worked hard over the years to keep our overhead in check, generally rising labor costs continued to drive up the price of doing the surveys for the school districts, especially since it required our personnel to travel all over the state to administer and collect them. In addition, the State of Michigan began to expand the services offered through the Office of Drug Control Policy (including alternative procedures for evaluating school drug education and prevention programs) to the point that our services were no longer needed, and therefore it was no longer necessary for our research center to continue the project. Nevertheless, for nearly two decades and for more than 85% of the public school districts in Michigan, MAOD provided a valuable service. We provided current, accurate data to hundreds of school districts that they used in the planning, design, implementation, and evaluation of their drug education and drug prevention programs.

Unfortunately, our data suggest that the end results of the War on Drugs in Michigan are not terribly encouraging. Tobacco and steroid use have clearly declined, but there has only been moderate change in alcohol use. Although marijuana use did drop in the late 80s and 90s, it started to climb again toward the beginning of this century. Moreover, there have been and continue to be small but persistent numbers of students who use—and abuse—prescription drugs, inhalants, and over-the-counter drugs, as well as even smaller numbers who use substances like meth, crack, and heroin.

Even though the overall results were not particularly encouraging, it is important to remember that without the project it would not be possible to draw those conclusions, nor would those school districts have been able to realistically carry out their programs. Carefully designed and executed survey research can tell us much about what people do and why they do it. After all, to describe, understand, and explain human behavior is what applied researchers do.

USING SOCIOLOGICAL SKILLS FOR MOVEMENTS TO CONFRONT POWER: THE GENESIS OF THE RESEARCH INSTITUTE ON SOCIAL AND ECONOMIC POLICY (RISEP)

Bruce Nissen

Center for Labor Research and Studies, Florida International University, Miami

Bruce Nissen is a sociologist and labor educator whose scholarly research has mostly centered on social movements, labor–community coalitions, and activist labor unions with which he has been associated. He previously taught at Indiana University and recently has been dividing his time between employment at the Center for Labor Research and Studies at Florida International University and employment as the staff representative for the higher education faculty union in Florida, the United Faculty of Florida.

It all started innocently enough. For about two decades, I had made a "scholarly career" out of studying and writing about the social movements and social causes around me. I had joined some of them as an individual because of my passionate interest in making the United States a more just society according to my standards of social justice. For me, being a labor sociologist has always meant researching and writing about the movements of ordinary working people to achieve more power and a fairer distribution of the wealth created by their work. Through these efforts, I found myself in Miami Beach, Florida, at the June 2003 national conference of the organization Jobs with Justice. I was attending as a delegate, since I was serving on the executive board of the local Miami chapter.

Seemingly out of nowhere, a conference attendee approached me and introduced herself as a representative of a small foundation that specializes in funding social movement organizations. She informed me that she had been impressed by the conference's evidence of an upsurge of labor and working-class community activism in South Florida. When she asked the activists she met at the conference how they handled the research piece so necessary for intelligent campaigns on behalf of working people, she told me that they were all answering, "Oh, Bruce Nissen from Florida International University does the research."

My response was incredulity: I told her that I was only one person, that I only did a limited amount of research compared to the need, and that I felt working people and their organizations needed much better research capabilities than that. She agreed and said,

> That's right, and that's exactly why I want you to set up an institute or center to carry on this kind of work on a bigger scale. If you would do that, my foundation would fund you. We're small, so we wouldn't be able to give you a lot, but I also will advocate for larger foundations to fund you more fully if you'll set it up.

She was part of a group of foundation program officers who dubbed themselves the "Social Justice Infrastructure Funding Group," and they had a particular interest in Florida, among other places.

The foundation representative's offer certainly piqued my interest. To make a long story short, by the fall of 2004, in conjunction with colleague Carol Stepick who had a great deal of experience in running research field projects, I set up the Research Institute on Social and Economic Policy (RISEP). We organized it as an unofficial research institute of Florida International University (FIU), with modest initial general operating support from the Rockefeller Foundation and a couple of much smaller funding sources.

From the very beginning, RISEP was set up along unique lines. It had to achieve two objectives if it was to fulfill my vision for it. First, it had to maintain standards of scholarly research that established it as a credible source of fact-based research shedding light on social institutions and social problems. Second, it needed to illuminate aspects of social reality that are often hidden or ignored because ordinary workers and the have-nots in our society have few resources or opportunities to bring them to light, compared to elites who routinely hire the best minds to help them maintain their wealth and control.

The solution to reconciling these two purposes was to (1) insist on total freedom as a scholarly research institution to arrive at conclusions as they appeared, free from spin; and (2) make sure that the *questions* we asked and investigated were those asked by people representing working-class communities and organizations engaged in bettering the lives of working people. Thus, RISEP does not allow anyone—funders, constituencies served, politicians, the university administration, or anyone else—to control the results of our research. However, it does orient itself to the questions of interest to activist organizations serving working people, and thus it researches questions that are unlikely to be researched by organizations serving the government or funded by mainstream sources.

RISEP's self-description on its Web page gives a good idea of its orientation:

> RISEP is a research institute that studies issues of concern to working people and low-income communities. We provide facts, figures, and information to labor unions, community organizing groups, and the media about things that impact the lives of working people: [T]he economy, wages and benefits, housing cost, health care, immigration and the environment are some issues we have studied. (www.risep-fiu.org/about/)

The RISEP Advisory Committee provides feedback on research reports, gives advice on what topics are most important to address in unsponsored research projects, and frequently offers sponsorship of targeted research addressing topics of concern in ongoing organizing and campaign work. Its composition again gives an indication of the types of constituencies that find RISEP's research to be most valuable. Members include representatives from (1) faith-based groups organizing for social justice, such as the South Florida Interfaith Worker Justice organization; (2) activist labor unions and labor bodies such as the South Florida AFL-CIO; (3) labor–community coalitions, such as the South Florida Jobs with Justice chapter; (4) community organizing groups, such as the Miami Workers Center and Power U Center for Social Change; (5) immigrant-based organizations, such as the Florida Immigrant Coalition and the Coalition of Immokalee Workers, which organizes tomato pickers; (6) human services providers, such as the Miami Coalition for the Homeless and the Human Services Coalition; and (7) progressive academics who bring university-based resources and insights.

Between the fall of 2004 and the fall of 2009, RISEP produced reports on affordable housing for low- and moderate-income neighborhoods; lack of health care locally and statewide; minority and women workers' issues; work, wage, and income conditions in the state; Florida's economic conditions; the treatment and status of immigrant workers in the state; labor and union issues; the consequences of requiring "community benefits" from developers and corporate recipients as a condition of receiving public subsidies; and a number of other topics, including analyses of the spending priorities of FIU itself. In total, the institute has produced 52 reports in that five-year period, many of them directly related to ongoing organizing work and campaigns for justice. All of these reports are available on the RISEP Web site. Often the results were published in the Florida news media; newspaper references to these reports average about 75 per year, and RISEP authors appear on radio and television shows an average of once every other month.

RISEP's research has helped the South Florida Jobs with Justice chapter win a community benefits agreement requiring that the construction of a new Florida Marlins stadium be built in a manner providing millions of dollars of benefits to local workers and communities. Likewise, it helped the Miami Workers Center win guarantees that affordable housing would be provided to the displaced public housing residents in the African American community of Liberty City. (This victory was later reneged upon by public officials, however.) RISEP research helped the Coalition of Immokalee Workers (CIW) expose as unscientific and biased a so-called "research report" commissioned by the McDonald's Corporation to discredit the CIW's demand for a "penny a pound" more for tomatoes picked and sold to McDonald's. It has helped a union representing workers at Disney World expose the negative consequences for the community of Disney's two-tier wage structure that provides lower wages for new employees than for others doing the same job. RISEP research played a key role in winning living-wage ordinances in several South Florida counties and municipalities. Many other examples could be cited as well.

RISEP also provides students at FIU an opportunity to be involved in research on real-world issues—research with real-world consequences. Sociology graduate students have gone on to pursue doctoral dissertation topics first explored through RISEP research. Also, with the exception of myself and cofounder Carol Stepick, paid RISEP personnel (which fluctuates between three and six, depending on funding and research project needs) have come entirely from students recently graduated from either master's or undergraduate programs. All these employees are pursuing further advanced degrees as they work at RISEP; their employment at FIU provides them with six hours of free tuition at the university per semester. Other students who have worked with RISEP have gone on to work with community organizing groups they first encountered through RISEP research. The institute has proven itself to be a natural home for students with a burning passion for social justice issues.

Readers are invited to explore the work of RISEP further through its Web site, www.risep-fiu.org. RISEP's real-world orientation directed at power-building organizing and progressive advocacy groups could be duplicated on other university campuses. I have been puzzled about why there are not more such university-based research organizations in existence, but funders and others tell us that we are fairly unique. I can certainly say that my life has been greatly enriched because of this unique institution. It has allowed me to build relationships with some of the most moral and visionary people I have ever met, and it has given me great satisfaction to see RISEP's research help organizers transform power relationships to the advantage of our society's have-nots. Perhaps spreading the word about RISEP will encourage others to attempt something similar.

POSITIVE DEVIANCE RESEARCH AS A WAY TO HELP PEOPLE

Samuel R. Friedman

National Development and Research Institutes, New York

Samuel R. Friedman is an author of 400+ publications on HIV, STI, and drug use epidemiology and prevention and about 50 publications on workers' movements, political economy, or social movements—including *Teamster Rank and File*. He has published two poetry chapbooks (*Murders Most Foul: Poems Against War by a World Trade Center Survivor*, Central Jersey Coalition Against Endless War; and *Needles, Drugs, and Defiance*, North American Syringe Exchange Network) and a poetry book (*Seeking to Make the World Anew: Poems of the Living Dialectic*, Hamilton Books). He regularly takes part in antiwar activities.

A large part of what sociological researchers study falls into one of two kinds of studies:

1. Research to deepen, prove, or disprove one or more sociological theories or hypotheses; or

2. "Social problems" studies of situations or behaviors that the researcher thinks are harmful, in order to attract attention to it or to figure out how to fix these harmful realities. Such "social problems" research also forms the basis of much sociological practice.

Some of my most useful research has been of a third kind called "positive deviance" research. It studies how some people or groups manage to create good (positive) outcomes in circumstances where most similarly situated people or groups suffer or do damage. The term *positive deviance research* was invented by international public health researchers who noticed that, in some very poor villages in which almost all the children suffered from severe malnutrition, a few equally impoverished kids avoided this and were in good health. They then studied what these children's parents were doing that produced these positive results.

Here, I will describe a positive deviance research project I conducted in the 1970s. I have conducted others since. All were very interesting projects to do, and all were also pleasant to do. In various ways, all have had real-world consequences that I think have improved the lives of many thousands of people. These studies have differed very much in content. In this study, I focused on local labor union activism.

Social movements often emerge in waves, as one inspires potential activists in other social domains to take mass action. In 1970, during the height of the radical black, Latino, women's liberation, student, and socialist movements of the 1960s and early 1970s, and of a more limited but strong upsurge of rank-and-file struggles by workers to improve their working conditions and safety, truck drivers in the Teamsters Union struck against their employers, in opposition to the will of the officers of the national union. This strike took place in many parts of the country but was hardest fought, and lasted longest, in Los Angeles, California. In large part, this was because the employers decided to use the opportunity to destroy the power of Local 208, a local branch of the Teamsters Union with about 6,000 members who worked at dozens of different "freight barns" in the area. They wanted to do this because Local 208 was very good at helping its members to defend their contractual rights under the negotiated contracts between the employers and the union. They also wanted to do this because 208's members, assisted by their local union, often engaged in semi-legal strikes or other forms of direct action to defend themselves against management's disciplinary actions or to extend their rights beyond those agreed to in the contract. In a dangerous job like local freight delivery, such "extra" rights as having more security against managers' pressure to drive and unload trucks more quickly could mean the difference between having or not having a traffic accident or injuring one's back by having to rush when lifting freight. Such workers' power, though limited, also could reduce the chances of severe psychological stress from being threatened or insulted by a manager.

At the time of the 1970 Teamsters Union strike, I was a sociology faculty member at UCLA and involved in socialist politics. Along with other faculty members and students from colleges and universities in the area, we provided what little help we could to their strike in 1970 and later to their battle to restore their power at work after the national union leadership removed the officers of Local 208 and installed a bureaucratic, struggle-averse set of new officers. By 1973, this effort led to new elections in which the new officers included many previously well-known leaders from the 1960s. As a faculty member, I was expected to do research. I saw an opportunity to study this local union, learn how it had managed to become a democratic, militant, and effective local within a union infamous for sweetheart deals with employers and for corruption, and then to try to make this knowledge available to other Teamsters and other workers in Los Angeles and throughout the United States, so that they could improve their lives. Without ever having heard of the term, I undertook an exercise in *positive deviance research* to see how these workers, unlike the vast majority of other workers, in the Teamsters and throughout the rest of the world, were able to build an organization that could fairly and effectively defend their interests.

I did this study using qualitative research methods. I went with local union officials when they had meetings with management over grievances. Often these involved situations where management had underpaid workers for overtime or had suspended or fired workers for various rule violations. To get some perspective on the differences between Local 208 and the practices of officials in other locals, I spent a day going with the officials of another freight local (in nearby Orange County) that was described to me by my friends in 208 as a relatively good local. When I did this, I saw many differences—the officials of 208, but not of the other local, always talked with the drivers at the workplace before going into meetings with management, and never met with management without some of the freight barn's stewards or workers present. They also were less likely to "give in" to management demands.

I also conducted many interviews with officials from the local, with working stewards[1] who were elected by the members in each freight barn (unlike in most Teamster locals where local officials appointed the stewards), and with other workers. My interviews with them covered such topics as how the union functioned, the joys and challenges of doing freight work, and how their work and union activities fit into the rest of their lives. Many of these conversations also delved into the history of Local 208 and the trials and triumphs of each freight barn and the local as a whole, from the late 1950s to the time of the interview.

This research involved a lot of hard work, but it was also a lot of fun. I met many very wonderful people, and had to think my way through many interesting questions of history and of sociological analysis to try to figure out what was going on.

What did I learn? Partly, I learned in much more detail things I already knew—that employers try to get workers to do dangerous things, that

---

[1]In the Teamsters, stewards were workers who served as the immediate face of the union at each trucking barn. They would also attempt to solve most problems with management as they arose, and only call in the union's paid business agents when necessary.

employers routinely violate contractual agreements if it will help them make money, and that workers who resist such efforts often put their jobs in jeopardy. Nevertheless, many workers fight back in spite of these dangers, and when workers fight back, they often win. I was surprised at the extent to which the smaller daily struggles by workers to defend their working conditions helped them fight effectively for democratic control of their own union. I discovered that the system of grievance handling through which union officials meet privately with management to resolve situations where workers say their rights have been violated is a central part of the mechanism by which the collective power of workers is fragmented, demobilized, and bureaucratized. Therefore, this process does much to create and perpetuate perceptions of unions as an outside, alien force that fails to defend workers' interests. I also learned many lessons about how workers at a workplace in which they are disorganized and at the employer's mercy can begin the struggle for rights and power. Finally, I realized more fully how workers can ally their localized islands of organization with other groups of workers to form powerful combinations that can help each other and, in time, many thousands of other workers.

As I was writing up my research—a process that took some years—I also worked with Teamster activists in Los Angeles and later in the New Jersey area to which I moved in 1974 to help organize a national organization of rank-and-file worker activists. The ideas I had developed were of some use to this effort—though I learned far more from the activists than they did from me. I am proud to have helped a little, but I also know that their ideas are the root source of their successes. (Indeed, this is a key part of what my research showed.) This organization, Teamsters for a Democratic Union (TDU), still exists and still helps workers defend themselves at the workplace and in struggles to democratize their unions. TDU members have been and are officers in many locals, and have been a major force in national Teamster politics and in fights with employers over national contracts as well. When their efforts are successful, TDU helps local members maintain control over their (TDU-supported) officers and also helps their officers to resist the forces of bureaucratization.

I embodied my conclusions in a book, *Teamster Rank and File* (Columbia University Press, 1982), and a number of articles. After some argument, I convinced them to publish 500 paperback copies of the book at the same time they released the hardback edition, which was priced too high for most workers to afford. I bought these 500 copies and arranged for some to be sold through rank-and-file militant organizations. I gave or sold the rest to worker activists whom I knew or met. These books and articles reached many worker activists—particularly through TDU, which sold the book at its various meetings and by word of mouth—and over the years many of them have told me that the book and its ideas have been useful to them in their struggles.

I have conducted a number of positive deviance studies since I studied Teamsters Local 208. Some of these have addressed HIV/AIDS among drug users and other populations. Early in the AIDS epidemic, I obtained a job studying HIV/AIDS issues that affected injection drug users. In one positive deviance project on this topic, I examined organizations of injection drug users—a group often thought of as incompetent slaves to their addiction—and how these organizations helped to lead the fight to prevent the spread of HIV/AIDS among drug users in several countries. A recent positive deviance study led me to develop a new research design to study a crucial question—how some long-term injection drug users managed to run their lives in ways that led to their avoiding infection both by hepatitis C (a disease that causes liver cancer) and HIV. (An abstract describing this research design can be accessed online at http://www.ncbi .nlm.nih.gov/pmc/articles/PMC3141294/). A younger colleague, Pedro Mateu-Gelabert, is now developing programs to teach these lessons to other drug users and to test scientifically whether these programs succeed in helping people avoid infection.

When done well, positive deviance research helps us to learn from people who have themselves found ways to solve problems, and then to make this knowledge available for others to apply and perhaps improve. My studies of this kind have probably helped save or improve tens of thousands of people's lives. Thus, I recommend this approach as an excellent—though often challenging—research method.

DISCUSSION QUESTIONS

1. Why do you think Thomas Van Valey's research about alcohol and other drug use among Michigan high-school students is important, even though the results were not the positive outcome hoped for by those running the War on Drugs? Why is it important for those conducting evaluation research (as well as other types of research) to follow the scientific method?

2. Were you surprised by any of Van Valey's findings? Why or why not? How do his research-based results compare to your own observations from your high-school experience? What would you suggest as possible solutions to decrease drug use in high schools?

3. What are the two objectives Bruce Nissen said RISEP had to achieve to fulfill his vision? Why is each important to the mission of the organization? Why do you think there are very few organizations like RISEP?

4. How does RISEP "transform power relationships to the advantage of our society's 'have-nots'"? Had you realized that sociological research could be useful in this

way? Does this information make you more interested in taking a sociological research methods course? Why or why not?

5. According to Sam Friedman, what is "positive deviance" and why does he believe it is important to study?

6. What did Friedman learn from his study of Teamster Local 208? Imagine you are active in a union. How might you use Friedman's findings in your work with your union?

7. Think of something you would like to change on your campus (perhaps the lack of parking, or bringing more local and organic food to your cafeteria, etc.). Imagine you are being asked to conduct evaluation research to assess the need for and feasibility of this change. Using information from at least two of the Sociologist in Action pieces in this chapter, describe how you might go about doing so. Do you think this would be an easy or a difficult task? Why?

8. Which of the research projects described in this chapter interested you the most? Why? Which one did you find least interesting? Why?

RESOURCES

The following Web sites will help you to further explore the topics discussed in this chapter:

ASA Section on Methodology	http://www.albany.edu/asam/
Association for Qualitative Research	http://www.aqr.org.au/
Evaluation and Methods	http://gsociology.icaap.org/methods/
Evaluating Internet Research Sources	http://www.virtualsalt.com/evalu8it.htm
Research Methodology and Statistics	http://www.sociosite.net/topics/research.php
Research Methods Tutorials	http://www.sociosite.net/topics/research.php
Survey Design	http://www.surveysystem.com/sdesign.htm

To find more resources on the topics covered in this chapter, please go to the Sociologists in Action Web site at **www.sagepub.com/korgensia2e.**

Chapter 4

Culture

Culture, which is composed of the values, norms, and artifacts of a society, impacts every facet of our lives. The values espoused by our society influence what we consider important and desirable. The norms are the rules and expectations for behavior that help us learn how to act appropriately. Finally, the artifacts are the tangible objects created by people from a particular culture that guide how we work, play, and carry out most aspects of our lives. Most people only recognize the power of culture when they travel to a society with obviously different values, norms, and artifacts. However, sociologists are always aware of the impact of culture. The Sociologists in Action in this chapter have used this knowledge to promote positive social change.

Juliet Schor has used her sociological knowledge of culture in myriad ways in her efforts to fulfill the commitments of sociology. In her Sociologist in Action piece, "Academic as Social Entrepreneur," she describes how she and some colleagues created the now-famous publishing house, South End Press, in order to disseminate the works of many of the most important social justice thinkers of our time. She encourages you and your classmates to take similar actions, noting that "as an undergraduate, you too can get together with your peers, follow your passion, and create an enduring legacy for social change." Schor has also worked to create the Center for Popular Economics, which gives voice to a social justice-oriented economic analysis, and the Center for a New American Dream, which helps "lead the country to a more sociologically and ecologically sustainable way of living . . . by focusing on changing the consumer culture." Through her work with these organizations, her teaching, public presentations, and widely read books, Schor has helped innumerable people recognize and begin to address the overemphasis on work and material acquisition in U.S. culture.

In the second piece in this chapter, an excerpt from "The Diary of a Mad Social Scientist," Corey Dolgon describes a typical day in his life as a Sociologist in Action, sharing how "over the years, [his] instincts as a community activist and organizer have meshed in exciting ways with [his] instincts as a sociologist." As the director of the Office of Community-Based Learning at Stonehill College, Dolgon spends his hours working to influence the campus culture, helping students and faculty realize the full value of the community neighboring the school, and creating a norm of faculty and student interaction with members of the larger community. In this essay, he describes some of the many interactions he participates in during just one (exhausting and exhilarating) day, to help bring about these goals.

In the final piece of this chapter, "Youth Culture, Identity, and Resistance," Nilda Flores-González and Michael Rodríguez-Muñiz describe their sociological work with youth activists of the Café Teatro Batey Urbano in Chicago, Illinois. Batey Urbano is a Puerto Rican/Latino youth-run and -operated cultural center that "quickly gained recognition as a hub for critical creative expression and community-based youth activism." Flores-González's collaborative research project with Rodríguez-Muñiz and other youth leaders from Batey provides a powerful "example of how youth—if given the opportunity and autonomy—can make a positive impact on their community," and how sociological research can be "transformative and empowering for participants and their communities."

ACADEMIC AS SOCIAL ENTREPRENEUR: CREATING ORGANIZATIONS FOR SOCIAL CHANGE

Juliet Schor

Boston College, Chestnut Hill, Massachusetts

Juliet Schor is a professor of sociology at Boston College. She was originally trained in economics and spent almost two decades teaching in the Harvard Economics Department and the Harvard Women's Studies program. She is a former Guggenheim Fellow and has worked as a consultant for the United Nations. Her newest books are *Plenitude: The New Economics of True Wealth* (2010) and *Consumerism and Its Discontents* (2011). Schor is currently working on a cross-national survey of "conscious consumers" who align their purchasing decisions with their ecological and social values.

Being invited to write this piece takes me back more than 30 years, to my student days, when I started my first social change organization, a publishing house. The year was 1977 and I had just enrolled in a graduate program in economics. Most of us, teachers and students alike, had a commitment to social justice and social change, but our program was pedagogically conventional. We were a group of highly educated, privileged young people who wanted to take the amazing things we were learning in the classroom beyond the "ivory tower," to make them accessible to a much wider range of people. And so, long before the term *entrepreneur* was on anyone's tongue, eight of us came together and started a publishing house. Our idea was to bring unconventional views forward. We believed the economic system was inherently unfair, and also prone to instability and collapse. (We were right about that!) We were also feminists, we thought the United States was acting as an imperialist in Southeast Asia and elsewhere, and we were concerned about the invisibility of working-class people. And we were operating during a time when big questions about the nature of our economy and society were being asked across society.

We wanted to publish books on these issues. We were also hoping to make enough money that we could use it to spread the word through other ventures—a magazine, speakers bureau, school, community center. (We were inexperienced enough not to realize that publishing is a small- or no-profit margin business, although most of these further efforts did eventually come to pass.)

We set up shop in a house in Boston's South End, named ourselves South End Press, bought a typesetting machine, and started producing books. We lived together in the house. We operated as a collective without a corporate hierarchy so that each of us got involved in all aspects of publishing, from typing to editing to publicity. We published many of the world's most important social justice writers, such as Noam Chomsky, Howard Zinn, Barbara Ehrenreich, bell hooks, Arundhati Roy, Vandana Shiva, Manning Marable, Cherrie Moraga, and Winona LaDuke. I suspect you have read some of these and other South End Press authors in your sociology classes. We brought out pioneering work on feminism, racism, and global capitalism; on propaganda and the military state; and on universities, the environment, and class in America.

I tell this story in some detail because it illustrates an important lesson about putting academic learning into action: It's never too early to start. I was just 20 years old when I began this project. I had barely begun my graduate education, so I wasn't ready to write my own books (that came later). But I chose an activity—publishing other people's work—that was appropriate to what I could do. And I believed in myself. We all did. We had confidence, a strong commitment to our mission, and we were willing to give our all to the endeavor. It wasn't easy, and we had some good luck along the way, but more than 30 years later, South End Press is still thriving. In the world of small presses, that's almost a miracle. As an undergraduate, you too can get together with your peers, follow your passion, and create an enduring legacy for social change.

Our publishing house also helped spawn another organization. As economics graduate students, we were critical of our professors for not doing more to bring their knowledge to a wider audience. This was even the case at the University of Massachusetts Amherst, a public university with a social mission. We students felt that the faculty was too oriented to other academics and professional success. The charges must have rung true, because they responded by starting an outreach organization of their own—the Center for Popular Economics. Along with one student (me), a group of faculty organized a summer program to teach our brand of social justice–oriented analysis to activists and people in the community. There was a need for this work because in the late 1970s, rich and powerful corporations started a national advertising and propaganda campaign, using major outlets, such as the *New York Times* and the Public Broadcasting System, to convince the country that there was no alternative to their program of tax cuts, less regulation, and larger government subsidies—for *them*. (Sound like a familiar message?) Ordinary people didn't have the tools or analysis to counter this onslaught of "free market" ideology. In an intensive two-week course, we shared our expert knowledge in an accessible, fun, and useful way. The Center for Popular Economics is still thriving, and has expanded beyond summer programs to create a range of workshops and a variety of materials. It is an example of combining scholarship and research in action to help people who are on the front lines of social change and social justice.

I received my PhD in 1982 and began my academic career teaching at Williams College and then Harvard University. I began to write my own books, hoping once again to take my message to an audience outside the university. In 1992, I published *The Overworked American,* which argued that a "work and spend" cycle was undermining the quality of life of Americans and degrading the natural environment. That book became a national best-seller. I followed it up in 1999 with *The Overspent American,* which revealed how social pressures and social inequalities led to a treadmill of spending, debt, and often unhappiness. (These themes burst into public consciousness during the 2008 economic meltdown.) Those books also led me into sociology, which had a richer tradition of studying consumer culture. It is also a discipline more tolerant of a diversity of opinion, a range of methods, and multiple theoretical approaches. One of the reasons I left economics is that, during the 20 years I was in that field, it became more oriented to supporting corporations and their interests, and less tolerant of diverse opinions.

By the mid-1990s, I was ready to start a third organization, and this time it was oriented more specifically to the analysis of our consumer culture and economy that I had been developing. I joined with a group of people from the nonprofit world (and one other academic), and we founded the Center for a New American Dream, a.k.a. newdream.org. Its mission is to help lead the country to a more sociologically and ecologically sustainable way of living, initially by focusing on changing the consumer culture. In the more than 10 years that we've been around, we've held contests for environmentally responsible behavior, persuaded companies to change their policies, and helped raise awareness about the impact of our lifestyles on the planet. We've also redirected billions of dollars in government spending out of environmentally destructive products, such as paper from virgin forests, toxic cleaning supplies, and computers that hog energy, into environmentally less damaging alternatives. As one of the founding board members, and current cochair of the board, I've been active in many ways, from conducting surveys of our members, to speaking on behalf of the organization, to blogging on our site and holding conferences. My academic background provides many ways for me to contribute to the work of the group.

Looking back on these three forays from the university into "action," I feel that they are among the most satisfying aspects of my 30-plus-year academic career. As early as high school, I wanted to use what I have learned in school to make change in the wider world. Sociology has been an ideal discipline for doing that. Many of its core concepts—inequalities, social exclusion, imperialism, economic exploitation, and human transformation of the natural environment—lead directly to a transformative agenda. Unlike economists and political scientists, sociologists have been willing to look at their own positions of privilege and analyze the conditions under which sociological knowledge is produced.

Most importantly, sociology is a field that not only accepts but thrives on diversity of thought. This keeps it relevant at a time like the present, when constant flux is the order of the day. My ideas have shifted enormously since I entered graduate school, and sociology appreciates such transformation. Intellectual flexibility has been vital to the success of all three organizations I've been involved with—South End Press, the Center for Popular Economics, and Center for a New American Dream—and I'd venture a guess that it's a core value for any successful endeavor in today's rapidly globalizing, changing world.

EXCERPT FROM "THE DIARY OF A MAD SOCIAL SCIENTIST"

Corey Dolgon

Stonehill College, Easton, Massachusetts

Corey Dolgon is a professor of sociology and director of the Office of Community-Based Learning at Stonehill College. He is author of the award-winning book, *The End of the Hamptons: Scenes From the Class Struggle in America's Paradise* (2005), as well as *Social Problems: A Service-Learning Approach* (2010, with Chris Baker), and an edited collection entitled *Pioneers in Public Sociology: 30 Years of Humanity and Society* (2010, with Mary Chayko). Dolgon is also a musician who has released three CDs and tours periodically with a singing lecture on folksongs and the U.S. labor movement called *In Search of One Big Union.*

8:30 A.M.: It's Monday morning. "My girls" are just out the door: Ruby (the 3-year-old) to preschool; Bailey (the 6-year-old) to first grade; and Deborah (my ageless wife who drops them off) to her consulting job coordinating a community overdose prevention coalition in Boston. By now I've made coffee, breakfasts, and lunches, and have showered, shaved, and packed my bag for work.

I am the director of the Office of Community-Based Learning (CBL) at Stonehill College, a small Catholic liberal arts school near Brockton, Massachusetts. I teach one class in sociology each semester, but most of my time is spent building the campus and community capacity for service learning and community-based research. Often this means late nights at community meetings or weekend visits to various events. Last night was one of those occasions.

Students and I attended a town meeting organized by the Brockton Interfaith Community (BIC), which is affiliated with People Improving Communities Through Organizing (PICO), a national group founded in 1972 and heavily influenced by Saul Alinsky's organizing theories and methods. My office linked some students with BIC to work on the housing foreclosure issue—Brockton has the highest rates of foreclosure in Massachusetts. PICO has recently been instrumental in pressuring the Federal Reserve to hold eight "town meetings" nationally to hear about the crisis directly from people losing their homes. Brockton was chosen as the final host for these events.

BIC organized a bus tour of the city and a round table for Federal Reserve officials to listen to local advocates and community leaders. But the heart of the evening was dedicated to a large public action where over 500 people turned out to hear local residents challenge not only Federal Reserve officials, but also state and national legislators, including Representative Barney Frank, head of the Congressional Bank and Finance Committee. Stonehill students had studied the issues, organized dozens of class presentations and events around campus, and led 25 of their cohort to join the event, and the students even organized some political theater to bring the issue to life.

Overall, the town meeting was highly successful: St. Patrick's Church was packed with people listening to friends and neighbors tell their stories and BIC leaders challenge Federal Reserve officials and members of Congress to adopt foreclosure prevention and new consumer protection policies. It was also very successful for the students. Many people met us after the event and said how much it had meant to have Stonehill students show support. Getting to participate—even in a relatively minor way—gave students a sense of ownership over the issue and a sense of belonging and solidarity with the community. One of the Stonehill students is now organizing a student chapter of BIC on campus.

Over the years, my instincts as a community activist and organizer have meshed in exciting ways with my instincts as a sociologist. Émile Durkheim (1947–1995) always helps me understand community and solidarity and the real need for physical participation in rituals as a form of identity construction. And Michael Omi and Howard Winant's (1994) racial formation theory informs my understanding of how racial and other identities evolve as movements organize and solidarity forms. This process impacts students' own analytical frameworks as well.

I note in my book, *The End of the Hamptons* (2005), that Southampton College students participating in a campaign to get outsourced custodial workers rehired were more likely to accept janitors' claims that racism was a key factor in the college's decision if they were regular attendees at meetings with custodians. Talking to and working side-by-side with custodians to discuss strategy, tactics, and—more importantly—the meaning of events and the organizing itself created a new collective identity for these students, an identity that recognized institutionalized racism at the same time that it challenged their own preconceived and rigid racial identities. Students supporting the custodians who only attended planning meetings with other students were more likely to believe administration claims that race had nothing to do with it. This second group of students, though still supportive of custodians' rehiring, never adapted their analytical frame despite strong evidence for racial discrimination. Similar to the Southampton College students who attended more meetings with janitors, the Stonehill students working alongside the good folks in Brockton were adapting solidarity through their hands-on organizing and personal connections.

9:05 A.M.: Hungover with exhaustion but also feeling a sense of triumph about the BIC event, I am a few minutes late for my meeting with three Stonehill faculty members. I have convinced my colleagues to deliver papers on their community-based learning course projects at an upcoming professional meeting, and two of the three are younger faculty I want to encourage to publish on their CBL teaching. Numerous journals, such as *Humanity and Society, Theory in Action,* and *The Michigan Journal of Community Service Learning,* now peer-review and publish such research, and I am hoping to develop a more integrative sense of civic engagement and scholarship among the faculty. In essence, I hope to place faculty in similar situations to the BIC students—thus reshaping their identity both as teachers and as scholars.

One sociologist is doing a community-based research project with her Sociology of Education seminar. She met with the superintendent of a local school district interested in better tracking of transfer students. This professor helped her students take the data and begin analyzing demographics and outcomes. The class also put together some qualitative interviews to explore students' own experiences and ideas.

Another sociology faculty member has students in his Social Problems class choose a service opportunity from the variety of projects sponsored by the college's Into the Streets service program. In a very traditional service-learning model, students work about two to three hours a week at local agencies and integrate course content on poverty, homelessness, and so forth with their field experiences. Students keep journals and write final papers on how experiences and readings influenced their thoughts on services and solutions. This professor and I discussed the possibility of developing more in-depth partnerships with just a few sites and integrating organizational staff and resources into the design of the course.

The final faculty member is a historian, and she has designed a learning community with another professor from the Visual and Performing Arts Department around Cape Verdean history and culture. Beginning last spring, the two faculty members met with Brockton's Cape Verdean Association (CVA) to start thinking about how Stonehill students could be matched with CVA youth in order to examine the history of the local Cape Verdean population. CVA youth and Stonehill students would then collect oral histories of people throughout the community and use them as the basis for a huge public art project. Meanwhile, the history professor created a Cape Verdean history course to give students more background on the particular history and culture of both the immigrants' origins and immigration patterns and experiences. Despite the different projects and approaches, each case is a powerful example of what students are gaining from integrating classroom discussions and readings with research, service, and creative organizing experiences in local communities. But we are also aware that the community impact of these projects is less guaranteed, and a focus on not just community input but community outcomes will remain a vital area for attention and assessment.

10:45 A.M.: I get into the office and sit down to answer my dozens of e-mail messages with some trepidation. It's all well and good for a Mad Social Scientist with a touch of Attention Deficit Disorder, but it makes for a very messy inbox as the e-mails pile up.

1:30 P.M.: I am rushing to class and thinking about how funny being a "sociologist in action" seems. I imagine there are few sociologists in "inaction." But to be *in action* must mean more than maintaining a frenetic pace of community meetings; faculty development; and traditional teaching, research, and publishing. For me, to be in action has always started from the theoretical framework of praxis: to study and learn while engaging community and its social problems; to research and write along with the people who know what questions to ask and how global forces impact their own communities; and to produce the kind of popular sociology that Gramsci (1972) argued was akin to his philosophy of praxis—where all people are philosophers engaging in critical and practical intellectual processes, in hopes of discerning "good sense" from "common sense."

4:15 P.M.: A new sociology faculty member comes by and asks if I have time to discuss his Social Movements class coming up next semester. He would like to incorporate CBL but is uncertain how.

We decide that we don't know much about the history of social movements in Brockton, so we map out a project that would have students choose organizations and interview leaders and members while also looking into some local history to understand the links between groups and traditional social movements. We will also help to organize events that might benefit some of these groups, such as a forum on environmental racism.

11:30 P.M.: Some days seem more frenetic than others. Some get swallowed up by logistics and administrative tasks and food shopping and endless e-mail. Some days there are inspiring conversations or events that reaffirm both individual and collective efforts and even promise that the world might one day be a better place. Some days I soothe my madness with a single malt before bed, finding an uneasy peace in the notion that I am doing my best by battling against pestilence, no matter how fragile or futile the effort may seem. But other days I unwind in solitude, swaddled by a sense of accomplishment, possibility, new solidarity, identities, ideas, projects, and people orbiting around me. My study of sociology has given me an understanding of the self's inherent social construction: We are inevitably a combination of what we do, what we say, what we write, and what we think, but all in connection with other people and their efforts, expressions, and ideas. The best work that we can do is in service to a vision of a more just, more humane, more dignified world for all. And every night, as I settle in, and I can hear my girls sleeping in the rooms around me, their breathing and dreaming suggests there might just be a method to this madness.

References

Durkheim, É. (1995). *Elementary forms of the religious life* (K. E. Fields, Trans.). Glencoe, IL: Free Press. (Original work published 1947)

Gramsci, A. (1972). *Selections from The Prison Notebooks* (Q. Hoare & G. Smith, Eds.). New York, NY: International Publishers.

Omi, M., & Winant, H. (1994). *Racial formation in the United States: From the 1960s to the 1990s* (2nd ed.). New York, NY: Routledge.

YOUTH CULTURE, IDENTITY, AND RESISTANCE: PARTICIPATORY ACTION RESEARCH IN A PUERTO RICAN BARRIO

Nilda Flores-González

University of Illinois at Chicago

Michael Rodríguez-Muñiz

Brown University, Providence, Rhode Island

Nilda Flores-Gonzalez is an associate professor in sociology and Latin American and Latino Studies at the University of Illinois at Chicago. She is the author of *School Kids, Street Kids: Identity Development in Latino Students* (2002). Her recent work on youth activism appears in her coedited book, *¡Marcha!: Latino Chicago and the Immigrant Rights Movement* (2010). She is also coeditor of *Immigrant Women Workers in the Neoliberal Age* (Forthcoming).

Michael Rodríguez-Muñiz is a PhD student in sociology at Brown University. He received his MA in sociology from the University of Illinois at Chicago. His research on Puerto Rican activism in Chicago's pro-immigrant rights movement appears in the volume *¡Marcha!: Latino Chicago and the Immigrant Rights Movement.* He is coeditor of a forthcoming special issue of *Qualitative Sociology* on new approaches to ethnography. His interests include culture, sociology of knowledge and science, social theory, ethnicity and race, and Latino politics.

I think hip-hop is who you are, what you live through. Hip-hop is a way of living, a way of life.

These candid words, spoken by a Puerto Rican youth in his early twenties, emerged from an innovative research project we embarked on with youth activists of the Café Teatro Batey Urbano in Chicago, Illinois. Though sociologists are expert analysts of urban life, they rarely conduct research *collaboratively* with members of the local community. Usually the researcher determines all major features of the study: what, where, when, and how. People in the community are viewed as potential sources of data and insight, but they are too rarely integrated into the design, implementation, analysis, and presentation of research. Hoping to counteract this, we instead undertook a Participatory Action Research study, an especially meaningful and satisfying form of research that develops from community input to the researcher. Our study investigated the use of hip-hop as a tool for liberation and, in doing so, contributed missing pieces to the ongoing sociological study of culture, identity, and community building.

Founded in March 2002 by a group of college students, Batey Urbano is a Puerto Rican/Latino youth-founded and -operated cultural center in the Chicago neighborhood of Humboldt Park. It quickly gained recognition as a hub for critical creative expression and community-based youth activism. During the time of our study, in 2004–2005, Batey was averaging over 150 events a year, attracting youth as young as 12 to watch and perform at hip-hop and poetry shows. Using urban youth culture as a form of resistance, Batey aims to raise awareness of the social forces shaping their collective experiences, as well as recognition of their ethnoracial, class, gender, and sexual identities. As more than just an "art for art's sake" venue, Batey encourages and empowers young people to struggle for self-determination and to reach out in solidarity to other marginalized communities similarly experiencing rampant poverty, racial exclusion, and educational disparities. Over the past eight years, the site has grown to offer writing and computer technology workshops; after-school programs; and training in online radio and music recording, theater, and activism in the ¡Humboldt Park NO SE VENDE! Campaign (Humboldt Park IS NOT FOR SALE!), which fights gentrification, a racialized process of high-end urban development that displaces long-standing residents and replaces them with people from social classes that can afford higher rents.

In the spirit of our collaborative research project, this Sociologist in Action piece narrates the experiences of the sociologist (Nilda) and one of Batey's founders (Michael), sharing lessons gained from the intimate and reflexive integration of sociological research and community activism.

Nilda Flores-Gonzalez:

When I was invited to contribute a chapter for a book on youth activism (Flores-González, Rodríguez, & Rodríguez-Muñiz, 2006), I immediately thought of Batey Urbano. I had seen them in various cultural events and in political actions, such as rallies and protests. Unlike predominant negative perceptions of inner-city youth, Batey is an example of how youth—if given the opportunity and autonomy—can make a positive impact on their community. Through my research and activist involvement in the Puerto Rican community since the mid-1990s, many Batey members knew me. Some knew me quite well, such as Matthew Rodríguez, who had been my student, and Michael, a recent college graduate whom I had worked with on several grassroots projects and was trying to recruit into graduate school. From the outset, I wanted the project to be a true collaboration, a partnership beneficial for everyone involved. I wanted to write an academic article, but one that would be meaningful for the community. My philosophy on research is one of engagement and inclusion of the community in the research process. To me, research should generate new knowledge, but it should also be socially relevant and transformative for the people we are studying. I also believe that as engaged sociologists we have to "give back" to the very community that has so graciously allowed us into their world. I saw this project as a way for Batey youth to share their stories and inform people about their creative community work. In the end, I think all of us gained much more than what we had initially envisioned. And that's the power of sociological research when done thoughtfully and respectfully.

Participatory action research is a dynamic research approach whereby community participants and the investigator work collaboratively to explore or address a specific social issue. It involves critical reflection of the entire research process, even the findings. Batey youth were involved in every step of the project: from gathering data to writing the article.

One youth member conducted interviews and recruited help from other members to transcribe them. Another member gathered materials (flyers, newspaper articles) and organized them into a binder. After I produced an initial draft of the chapter, the Batey youth gave feedback and edits, which I then incorporated into a new draft.

During the course of the collaboration, I had to reflect on the roles that I played as an outsider, community activist, and professor, as well as the age and class differences between Batey youth and myself. Although I am Puerto Rican, as are most of the members of Batey, I am a middle-aged and middle-class professor who is not into hip-hop. However, the fact that I am a *familiar outsider* with years of involvement in the local community and close relationships with some of the youth and their adult allies was invaluable for the project. In working collaboratively, I had to put to the side my role as a "professor" and listen to the youth's critique of my analysis and writing, incorporate their suggestions, and accept their line-by-line editing of the chapter. In the end, we produced a chapter everyone owned and felt proud of (and I succeeded in recruiting Michael into our graduate program in sociology!).

Michael Rodríguez-Muñiz:

When Nilda approached us with the idea of a research project on Batey, we were extremely excited by the proposition. As one of the founders, I felt that the research project marked an important milestone for our organization. It represented a validation of our efforts to create a cultural space of critical reflection and action, whereby young people in our *barrio* could intervene in securing a better future for our community.

Thrilled by the opportunity, we enthusiastically agreed, though I am not sure if we really understood what a collaborative project entailed. Only one of our members (Matthew), who became one of the cowriters of the chapter, had any research experience. As members of a community that has been pejoratively depicted by the popular media, journalists, and academics for decades, we were given the chance to help create and shape a scholarly account of our work and community. In this special case, voices traditionally silenced in narratives of urban life helped to expand existing knowledge of how young people imaginatively grapple with the realities of violence, drug addiction, poverty, and poor schools.

As the project unfolded, moving from interviews and discussions to reviews of writing drafts, we took the opportunity to reflect critically on our ideas and practices. I found this very refreshing because intensive and ongoing community work sometimes makes it difficult to reflect systematically on what one is doing. Despite our often slow response or unavailability for interviews, Nilda was determined to proceed jointly throughout the research. I recall Nilda's positive attitude throughout the project, even during the revisions phase. Members of Batey were charged with scrutinizing every draft, a task we took especially seriously. She welcomed our feedback and dialogued with us on the framing and organization of the chapter.

Personally, this experience sparked my interest in sociology and ultimately my desire and decision to pursue graduate studies. Through Nilda's principled engagement with us, I realized the significance of the *how* of research. While the outcome of the project was an achievement, I believe the process was by far the most educational and meaningful aspect. Far too often, scholars come to study people, especially those in subordinate social positions (e.g., women, lesbians, immigrants, and members of other groups not in the power majority), rarely viewing those studied as knowledge-makers. Our honest and open collaboration created an incredible synergy, which provides evidence of the fact that urban communities are, indeed, intellectual spaces.

Conclusion

Sociologists in action have the opportunity and responsibility to create the space for marginalized individuals and communities to share their histories, experiences, and lessons. There are countless ways that engaged sociological research actually impacts the real world. Though these experiences may not immediately or necessarily change social policy—even if that might be the broader intention—they are nonetheless transformative and empowering for participants and their communities.

After we made several presentations at conferences and local universities, Batey youth were invited to present their community work at a number of youth-focused gatherings and academic conferences throughout the country. The first of these events, held in Tennessee, was organized with the goal of drafting a national youth "Bill of Rights." Batey's presentation included the results of a survey they conducted in an alternative school some of them attended, a documentary on the effects of gentrification on their barrio, and the performance of several provocative spoken-word pieces. At another event, Batey youth traveled to Washington, DC, to present on their cultural programming and youth activism at an anthropology conference. Batey also hosted visits to their storefront site by various groups attending conferences in Chicago—including scholars and a group of youth activists from Tucson, Arizona. At these events and others, Batey leaders expanded their network, establishing relationships that in some cases grew into intimate exchanges between them and other youth activists. In their community, the book chapter they helped write has been used in educational forums, youth organizing trainings, and even grant proposals and other fundraising initiatives. Batey also takes pride in the fact that, since the publication of the chapter, several youth organizations modeled after Batey's critical philosophy and collective organizational structure have been created.

On their humble stage, within a small storefront youth space on the northwest side of the Windy City, Batey youth creatively cultivated their "sociological imagination" and applied it to make sense of their everyday lives and to refine their political practice. They devised ways to use cultural forms like hip-hop and poetry to connect individual, biographical memories to collective histories of ethnoracial exclusion, marginalization, and resistance. And by fusing sociological research and community building, a sociology professor and a collective of empowered youth activists realized the ideals of public sociology.

References

Flores-González, N. (2002). *School kids, street kids: Identity development in Latino students.* New York, NY: Teachers College Press.

Flores-González, N., Rodríguez, M., & Rodríguez-Muñiz, M. (2006). From hip-hop to humanization: Batey Urbano as a space for Latino youth culture and community action. In S. Ginwright, P. Noguera, & J. Cammarota (Eds.), *Beyond resistance! Youth activism and community change: New democratic possibilities for practice and policy for America's youth* (pp. 175–195). New York, NY: Routledge.

Pallares, A., & Flores-Gonzalez, N. (2010). *¡Marcha! Latino Chicago and the immigrant rights movement.* Urbana: University of Illinois Press.

DISCUSSION QUESTIONS

1. What prompted Juliet Schor and her friends to start the South End Press? What do you think might draw other people from relatively privileged backgrounds into social justice work? What, if anything, about your own background might propel you into similar efforts? Why?

2. Schor points out that the themes of her books *The Overworked American* and *The Overspent American* "burst into public consciousness during the 2008 economic meltdown." Why do you think people were so drawn to these ideas then? Do you think they should still interest people today? Why or why not?

3. What is your reaction to reading Corey Dolgon's Sociologist in Action piece? Does it make you more exhausted or more inspired (or both)? Why?

4. What does Dolgon mean when he says, "My study of sociology has given me an understanding of the self's inherent social construction: We are inevitably a combination of what we do, what we say, what we write, and what we think, but all in connection with other people and their efforts, expressions, and ideas." How does it make you feel about how you influence others and how others influence you? How can this be a powerful tool for you in creating social change?

5. Nilda Flores-González and Michael Rodríguez-Muñiz maintain that "sociologists in action have the opportunity and responsibility to create the space for marginalized individuals and communities to share their histories, experiences, and lessons." How does Flores-González's collaborative research with Batey exemplify this opportunity and this responsibility?

6. If you were a professional sociologist, do you think you would be willing to have subjects of your study conduct "line-by-line" editing of your work? Why or why not? Why do you think it was important for the success of Flores-González's research about Batey, and what do you think it added to it?

7. Imagine you are a Sociologist in Action. The president of your college has asked you to conduct a study on the campus culture and to then use those findings to help improve it. Describe how you might use what you learned from at least two of the Sociologist in Action pieces in this chapter to help you figure out how you might proceed.

8. Using information from at least two of the pieces in this chapter, describe how attempts to change a culture must involve a great deal of (a) social interaction and (b) patience.

RESOURCES

The following Web sites will help you to further explore the topics discussed in this chapter:

ASA Section on Culture	http://www.ibiblio.org/culture/
Cultural Studies and Theory	www.media-studies.ca/resources/cult.htm
Social Movements & Culture	http://culturalpolitics.net/social_movements
Sociosite—Culture	http://www.sociosite.net/topics/culture.php
What Is Culture?	cooley.libarts.wsu.edu/garina/soc101/ Documents/Lecture_5_Culture.ppt

To find more resources on the topics covered in this chapter, please go to the Sociologists in Action Web site at **www.sagepub.com/korgensia2e.**

Chapter 5

Socialization

We learn how to interact effectively in society through socialization. Through the socialization process, we learn the norms and values that our society and socializing agents deem to be important. We also learn where we fit into our society and who we are as individuals. In order to become socialized, we must interact with other people. Of course, some of our interactions have more of an influence on our socialization process than do others. Those people who have the most influence over us are called *primary socializing agents*. Those who have some influence over us, but not as significantly, are called *secondary socializing agents*. In this chapter, we will look at how sociologists are using their knowledge of socialization to improve the experience of students taking international service trips, to create avenues of interaction between students and people who are homeless, and to improve interactions between youth and police. All of these are shining examples of how sociology can be used to help young people make society more just.

In "Socialization, Stereotypes, and Homelessness," Michele Wakin relays some of the opportunities she gives students to interact with homeless Americans, such as a project through which they meet and interview homeless people. As Wakin notes, "Through participating in this project, students begin to question their own role in creating social change by becoming involved in the lives of others less fortunate than themselves." Wakin also vividly describes the powerful impact on her students of participating in the annual Point-in-Time head count of homeless people in their surrounding community. Students come into direct contact with homeless women, men, and children living on the street (and in the woods) and in shelters. These interactions help "debunk the stereotypes associated with homelessness by tapping into the firsthand knowledge of people experiencing it" and spur many of her students to take action to curb homelessness.

Shelley White illustrates how discovering a missing element in the international service trip (IST) experiences at her college helped her and her colleagues realize why levels of activism were low on campus. In "Reengaging Activism in the Socialization of Undergraduate Students," she describes how "service programs abound on this campus, while opportunities for learning about activism and structural change are sparse." Examining the content of IST experiences, White and her colleagues "found that there, too, service was more reinforced than activism as an appropriate avenue for engagement." She was able to use this finding to create opportunities for students to learn about the potential power of social action to effectively confront inequality. Her piece is a remarkably clear and inspiring example of fulfilling the core commitment of sociology: to use the sociological eye to notice social patterns and then to use the knowledge gained by sociological research to make a positive impact on society.

Susan Guarino-Ghezzi closes this chapter with "Dangerous Behaviors? Police Encounters With Juvenile Gang Offenders," a dramatic illustration of how she used sociological tools to discover that police and juvenile offenders were each "locked . . . into routine, ritualized behaviors, guaranteeing that they would clash." Her "goal was to uncover these patterns, expose untrue and misleading stereotypes, and to use these new understandings to change behaviors on both sides." Guarino-Ghezzi shows how she fulfilled this goal, changed the norms of interactions between both groups, and made an important breakthrough in efforts to understand how juvenile crime can be reduced.

SOCIALIZATION, STEREOTYPES, AND HOMELESSNESS

Michele Wakin

Bridgewater State University, Bridgewater, Massachusetts

Michele Wakin is Executive Assistant to the President and chair of the President's Task Force to End Homelessness at Bridgewater State University in Massachusetts. Her research focus is on how marginalized communities survive and enact resistance in urban spaces. She received her PhD at the University of California, Santa Barbara, and wrote her dissertation about vehicle living as a form of homelessness. She has articles in *American Behavioral Scientist, City & Community, Administrative Theory and Praxis,* and *Journal of Workplace Rights.*

Sociologists view socialization as the process of developing an understanding of ourselves and our place in society through social interaction. Our socialization process also influences how we view various groups
in our society. Socializing agents (such as our families, the media, and our
peers) guide how we view different groups of people and lead us to judge
members of these groups in certain ways. Homeless Americans are one
group of people generally perceived as unable or unwilling to follow many
of the dominant norms of our society (such as providing shelter for oneself,
working hard, etc.).

Examining the accuracy of the prevailing view of homeless people
requires critically exploring the stereotypes that surround homelessness.
Stereotypes are used to describe categories of people who have not "made
it" and often attribute to them negative characteristics. If we hold the general belief, for example, that homelessness can be avoided with enough hard
work and perseverance, we also believe that an individual who is homeless
is homeless by choice or because of personal failure. In other words, we
tend to apply the stereotype *both* to the group as a whole *and* to each individual member of the group.

In order to illustrate how socialization and stereotypes work, I require
students in my Homelessness in U.S. Society class to engage in two civic
engagement projects. The first project is designed to bring students into
direct contact with homeless people in order to challenge prevailing stereotypes, to examine people's pathways into homelessness, and to explore
resources needed to escape it. The second project includes both quantitative
and qualitative components, as it involves a night count of homeless people
and a series of interviews. Both parts of this project are designed to show
students how different types of data can affect social policy.

In preparation for the first project, I ask students to close their eyes and
picture a homeless person. The majority of the class generally pictures an
older man with a scraggly beard pushing a shopping cart. I then ask them
to write a brief description of this hypothetical man's life. Most students
imagine him having alcohol or drug problems and mental health issues. We
then examine where these ideas about homeless people come from. Many
students cite media images or individuals they have seen on the streets during trips to Boston or New York. It is a surprise for them to learn that the
fastest growing segment of the homeless population is children and families
(National Coalition for the Homeless, 2009; Shinn & Weitzman, 1996). To
illustrate how children are affected by unstable housing, we read the book
There Are No Children Here (Kotlowitz, 1992), which details the lives of
two children growing up with their family in the Henry Horner housing
projects in Chicago. Although they are not homeless, growing up in housing projects where one's home is not a safe place exposes these children

to various risks and illegal activities. This book illustrates the crucial role that housing plays in determining everyday opportunities as well as overall life chances. It also connects the idea of unequal access to resources with real-life examples of the struggle for survival in a low-income community. Given the impact that the current housing crisis has had on the number of homeless children and given the complex challenges they face, this is a particularly timely issue to discuss (Duffield & Lovell, 2008).

With this information as background, I arrange a field trip to a local emergency shelter, where my students tour the facility and meet with homeless people residing in the shelter. Prior to our visit, students write a series of questions that are distributed in advance. Some of the questions past students have asked include the following: Did you think you were going to be homeless when you were young? Do you have any family that you could turn to if you wanted help? Have you seen people's (family/friends/strangers/employers) view of you change since you became homeless? If yes, how?

Last semester, three homeless men and one homeless woman agreed to meet with us and answer our questions. Our discussion focused on domestic violence and posttraumatic stress disorder (PTSD) among veterans as pathways into homelessness. Students were surprised to hear that homelessness was not always a lifelong condition, but was often a temporary lack of housing due to tragic events. They were also surprised to hear that all but one of the people we spoke with remained in contact with their immediate family. The homeless individuals we met with spoke candidly about how difficult it is to be seen as homeless and said that it was not uncommon for former friends to turn their backs. They also indicated that limited access to resources such as training, education, and affordable housing were barriers to gaining employment, leaving the shelter, and becoming housed.

Overall, this activity broadens students' understanding of homelessness and awakens a desire to participate in community service activities. As student Amy Cavanaugh writes, "As a student in this course, my mind has been opened. Now I participate in as many community service and outreach programs as I can. I never understood the reality of homelessness before, [only what I saw in the news]." Student Britney Garfield also expresses a sense of social responsibility: "I now feel it is my responsibility to help those who are suffering from homelessness and pass on knowledge of their suffering and neglect to others." Through participating in this project, students begin to question their own role in creating social change by becoming involved in the lives of others less fortunate than themselves. After graduating, student Jillian Miceli went on to make this issue a part of her professional life by becoming the Program Coordinator for Horizons for Homeless Children, southeast region.

The second project corresponded with the annual Point-in-Time count of homeless people in Brockton, Massachusetts. The count is part of an annual funding application to the U.S. Department of Housing and Urban Development (HUD) and is conducted nationwide by all communities receiving federal funding. The three basic components of the count are (1) an inventory of shelter beds available, (2) an inventory of those occupying the beds, and (3) a count of those who are unsheltered. The count occurs at night during the last week of January, when shelter usage is likely to be highest. Students from my course participate in the unsheltered count, which means spending several hours driving from location to location in search of unsheltered homeless people. Last year, there was a snowstorm on the night of the count, but to our surprise, we found a man sleeping in a tent behind a local mall. We made noise and yelled "hello" as we approached. The man came out of his tent, greeted us, and showed us the features of his living area. It was not only surprising to find someone camping out on such a cold night, but the man's articulateness, patience, and candor also were illustrative of the injustice of many stereotypes about homeless people. Student John Kennedy was struck by what he saw: "This man was intelligent. He was a homeless veteran . . . and his world was the camp he had built in the woods and now he was just trying to survive with a little dignity."

Students were also impressed with the idea that our count would shape the direction of future homeless services by offering a numerical estimate that could be used in comparison with other years and in demonstrating the need for additional housing and shelter. As senior Justin Mitchell wrote, "We spent as much time *making* social change, through the homeless count, as we did studying the social problem" (emphasis original). Student Dan Kent concurs:

> The homeless count left an impact on me that made a semester of studying homelessness feel real and life-changing . . . the homeless count felt more real than anything I could have imagined. It was life beyond the textbook that most students will never see.

To assist students in further connecting their interest in homelessness with regional service provision and policy, I applied for a Community Action Research Initiative grant through the American Sociological Association. The grant required a partnership between college and community organizations and a focus on social justice. The purpose of the grant was to explore the feasibility of providing a qualitative component to the annual Point-in-Time count. Bridgewater State University students worked with local shelter providers and the Plymouth County Housing Alliance to gather extensive demographic data on the local homeless population, an important preliminary step in identifying the parameters of our sample. Students were awarded work-study compensation and participated in an awareness training to prepare them to conduct qualitative interviews in Plymouth and Brockton. Each interview lasted approximately 15 minutes and explored factors leading up to homelessness, homeless services, and past and future housing alternatives. Each person who agreed to be interviewed received a $10 gift card for his or her participation. We collected 39 interviews in all and presented our results at the annual conference of the National Alliance to End Homelessness and to the Plymouth County Housing Alliance.

In the future, we plan to include qualitative data as a regular feature of the Point-in-Time count and make our findings available for future funding requests to HUD and to the Massachusetts Interagency Council on Housing and Homelessness. This experience was another way for students to connect their classroom learning with social policy and social inequality. Student Jason Desrosier sums up his experience this way: "The interviews were a unique and empowering experience in which the direct needs of homeless individuals were addressed. The interviews were an important step in the right direction to overcome homelessness and effect social change."

Students participating in these projects as a feature of the course Homelessness in U.S. Society critically examined socialization and stereotypes using quantitative and qualitative methodologies. The interviews they conducted, as well as the shelter visit and Point-in-Time count, brought them into direct contact with homeless people in both shelter and street settings. This helped debunk the stereotypes associated with homelessness by tapping into the firsthand knowledge of people experiencing it. The demographic interviews built on the initial shelter visit and were an important addition to the street count. They allowed students to capture more detailed information on the personal backgrounds of homeless people, reasons they were without housing, and their most pressing service needs. Perhaps most importantly, these projects inspired students to become agents for social change and to use their newfound knowledge and methodological tools to work toward righting the wrongs of class inequality.

References

Duffield, B., & Lovell, P. (2008). *The economic crisis hits home: The unfolding increase in child and youth homelessness.* National Association for the Education of Homeless Children and Youth. Retrieved from http://www.naehcy.org/econ_down_rep.htm

Kotlowitz, A. (1992). *There are no children here: The story of two boys growing up in the other America*. New York, NY: Anchor Books.

National Coalition for the Homeless. (2009). *Fact sheet: Homeless families with children*. Retrieved from http://www.nationalhomeless.org/factsheets/families.html

Shinn, M., & Weitzman, B. C. (1996). Homeless families are different. In J. Baumohl (Ed.), *Homelessness in America* (pp. 109–122). Phoenix, AZ: Oryx Press.

REENGAGING ACTIVISM IN THE SOCIALIZATION OF UNDERGRADUATE STUDENTS

Shelley White

Worcester State University, Worcester, Massachusetts

Shelley White is Assistant Professor of Public Health at Worcester State University. She received her PhD in Sociology from Boston College and holds a master's degree in international public health. Shelley has worked in HIV/AIDS policy and programming with the Maine Department of Health and Human Services, and the Ministry of Health in Lesotho, Southern Africa. In addition, she works with several international organizations on action planning for social change, including Free The Children, the People's Health Movement, Me to We, and SocMed. Shelley teaches and has published on such topics as HIV/AIDS and public health, globalization and political economy, human rights, development, and social change.

My path into sociology began through a career in health care, and my orientation to activist scholarship was built gradually through years of service engagement. Each of these transitions was facilitated by powerful experiences that exposed me not just to the deep inequalities that exist in the world, but also to the political, economic, and social structures that continually reproduce these inequalities (such as immigration policy, global trade and lending policies, racism and discrimination, food and agricultural policies, corporate tax structures, etc.). I think I also made these transitions because key experiences and role models taught me that, indeed, one can make a positive impact on the world!

As an undergraduate student at Boston University, I traveled on an international service trip (IST) just over the U.S. border to Tijuana, Mexico.

Although my group was working only a few miles over the border, we were exposed to a whole new world just south of the United States. We lived and volunteered at a migrant shelter for youth, and though most of our work was physical—replacing a roof, building closets, doing demolition to create new bathrooms—I think the major work we came for was emotional and educational. We had the chance to meet young people separated from family members across the border by a harsh corrugated metal wall, struggling with poverty and bleak future prospects in their country, or escaping conditions of child labor and other abuses. We visited a variety of ad-hoc homeless migrant camps, shelters, social service programs, and even a street theater program for child prostitutes and children engaged in other night labor in service to U.S. tourists. While we learned about people's experiences, we also learned about the history of U.S.–Mexico border relations and various iterations of U.S. immigration policy.

For me, this particular experience was so personally transformative because, while we engaged in service, which felt meaningful and fulfilled critical needs, we also learned about the structural causes behind the conditions of poverty and inequality we observed, and we learned about efforts—past and ongoing—for taking action and making more permanent change on these issues. Our learning included dialogues with immigration activists who taught us about their powerful approaches to change making. For me, this IST experience began my own socialization process in understanding my connection to global issues as a U.S. citizen, and my responsibility to engage in informed action for social change.

Years later, after completing my master's degree in international public health, and working in health policy and programming in the United States and southern Africa, I decided to complete my PhD in sociology. I entered my doctoral studies at Boston College (BC) with a great cohort of students, many of whom were interested in understanding how sociology could lead to sustainable social change, and we came together to form a graduate student "public sociology collective." Although we were surrounded by amazing and inspirational public sociologists, we didn't have formal mechanisms for learning the theory and practice of public sociology. We approached our department's faculty in an open-forum meeting about our desire for more opportunities to learn about and practice public sociology, and they agreed to create a two-semester practicum course. Through the Public Sociology course, we read about and discussed concepts of activist scholarship and learned about how sociologists use sociological methods and perspectives to create structural social change. The other major piece of the course was a research project about civic engagement on our college campus.

In our dialogues about our own activist and scholar identities, and more broadly, the role of academic institutions in facilitating learning and engagement for social change, we also turned our sights to undergraduate experiences of civic engagement at BC. We were struck by the social importance of ISTs on our campus; from conversations with undergraduates, we knew that students competed heartily to go on ISTs, and many seemed to speak about service trips as a sort of badge of honor. At the same time, given the types of experiences I and others in our collective had had on ISTs, we were surprised to learn that activism was quite stigmatized at BC. We heard from student activists that they felt marginalized on campus, several stating that the general student body seemed to consider activism a "dirty word."

For the research project for our public sociology course, we decided to study the role of ISTs in shaping students' understandings of service and activism, and their inclination to engage in either form of civic engagement upon return. We wondered what part ISTs, as a very visible and coveted form of service, were playing in contributing to students' socialization around service and activism. We considered service to be an approach that fills an immediate need (i.e., feeding a person who is hungry), but does not upset existing power differentials between the server and the receiver and does not aim to solve the social problem. Activism, in contrast, aims to address the existing power differential and to resolve the social problem more permanently, usually at structural or policy levels. To explore this question, we completed interviews with participants from several recent ISTs, asking them about the IST, their reflections on social problems and solutions, and their thoughts on service and activism.

Our research revealed several findings. First, students seemed to return from ISTs with a deep sense of dissonance—that is, they were deeply moved by their experiences abroad, but this was coupled with great uncertainty about *what to do*. Second, based on their own definitions, students tended to uncritically valorize service as a mode of civic engagement, but expressed ambivalent and negative feelings about activism. When students were asked to define and compare the two forms of engagement, one theme that struck us was students' reflections that service is ultimately a more *available* form of engagement than activism, and that even if students were interested in activism, they were not sure how to begin engaging in activist work (Cermak et al., 2007).

The latter finding, in particular, led us to what seemed a critical piece of the picture—students' vague and negative conceptions of activism may be due at least in part to their lack of exposure to activism and its potential. While students did reflect on the importance of sustainable change in their discussions of social problems and solutions, they seemed quite uncertain of *how* one actually enacts structural solutions. Service programs abound on BC's campus, while opportunities for learning about activism and structural change are sparse. In probing the content of ISTs, we found that there, too, service was the paradigm being reinforced as the appropriate avenue for engagement.

Our group presented the findings of our research multiple times on campus to many of our publics, including administrators of service trips and service programs. Fortunately, our findings were well received. Since then, we have been invited to run workshops on campus for two prominent service learning programs, in which we teach student-centered modules about the history, ideology, and tools of activism. Over the past several years since we completed our study, I have also consulted with the administrator who runs one of the college's largest IST programs. For this program, I now make a yearly presentation on globalization and the broad political and economic factors that help explain poverty and inequality today to all IST participants. These types of discussions and teaching opportunities use sociology to situate the social problems students observe internationally in a structural analysis, and include many examples of activist movements working toward change. As one student commented,

> The lectures got our wheels turning on a lot of social issues. People came into this program with a desire for social change, social justice, and wanting to help people, but with really vague ideas and lacking direction. The pre-trip lectures were HUGE—they generated discussions and set the stage for what we saw on the trip.

Finally, and perhaps most importantly, this year, the large IST program implemented modules for action planning. All student leaders participated in a training module I designed about the steps for planning effective actions for social change. They then facilitated "solidarity projects" in which students translated their trip experience into an action following the trip. One group traveled to Guatemala and spent time with a small coffee-growing community of former guerillas. During their trip, they learned about Guatemala's civil war and the ongoing Truth Commission seeking justice for survivors of human rights atrocities. Upon return, they partnered with another local university to put together several events, including a documentary screening about this community and its struggles, a panel on Guatemala's human rights situation (which included student leaders as panelists), and events to support the local coffee production of this community. Another group visited Mexico and learned about Mayan traditions of community gardening and has since engaged in a Real Food movement on campus, supporting community and organic farming movements. A third group visited El Salvador and commemorated the 20-year anniversary of Archbishop Oscar Romero's assassination with a panel event that included Massachusetts Representative Jim McGovern. In the 1990s, McGovern helped expose the ties between the United States and the members of El Salvador's military responsible for murdering Jesuit priests in 1989, and led the effort to cut U.S. aid to the country's military and bring about peace accords. Representative McGovern's example of activism was instructive, on this anniversary, of the power of global political mobilization.

Not all of the solidarity projects were activist in nature: Some raised funds; many raised awareness; and some campaigned for structural change. However, what the projects provided almost universally was an avenue through which students could avoid feelings of dissonance and translate their newfound passion about social issues into tangible actions with tangible results. According to one student leader,

> I was so excited to empower my participants to come up with a project and follow through. . . . It took a lot of time and effort, but it was 150% worth it! This is what it's all about. What's the point of going on a trip, raising $2500 to travel, if we don't *do something* about the social issues we've observed? (emphasis original)

Students learned concrete skills as they planned, problem solved, and ultimately carried out successful events—skills that should stay with them in their journey forward as change makers. The incorporation of action planning adds an important social change element to the socialization process students go through during their IST experience.

One of the gifts that sociology has brought to me is that of a broadened lens, one that has allowed me to understand the importance of both serving people's immediate needs and working to change the conditions that perpetuate their need. This broadened analysis began with my own IST experience many years ago, a journey that was infused with sociological analysis and that has guided my scholarship and activism since. Having had the opportunity to bring sociological analysis and activist learning into IST programming at BC in recent years has been an amazing experience, one that I hope will create an opening for students to understand their potential to effect positive social change in the world today. These experiences have reinforced for me that informed activism is a real skill, one that must be learned and imparted. They have also helped me to realize that teaching activist skills is tremendously rewarding and important, and at the very heart of the core commitment of sociology.

Reference

Cermak, M. J., Christiansen, J. A., Finnegan, A. C., Gleeson, A. P., Leach, D. K., & White, S. K. (2010). *Displacing activism? The impact of international service trips on understandings of social change.* Manuscript submitted for publication.

DANGEROUS BEHAVIORS? POLICE ENCOUNTERS WITH JUVENILE GANG OFFENDERS

Susan Guarino-Ghezzi

Stonehill College, Easton, Massachusetts

Susan Guarino-Ghezzi is a professor of sociology and criminology at Stonehill College. She is the former Director of Research of the Massachusetts Department of Youth Services, and is a frequent consultant on juvenile crime. She was principal investigator for research projects on juvenile sex offenders, ex-offender reintegration, deinstitutionalization, and staff–youth interactions. Her work has been cited by Amnesty International, the U.S. Department of Justice, and the PBS program *Frontline*. Professor Guarino-Ghezzi is coauthor of the books *Balancing Juvenile Justice* (2nd ed., 2005, with Edward Loughran) and *Understanding Crime: A Multidisciplinary Approach* (2006, with A. Javier Treviño). She has a PhD in sociology from Boston College.

If [people] define situations as real, they are real in their consequences.

—W. I. Thomas

W. I. Thomas was a sociologist who studied groups of immigrants who came to the United States during the early 1900s. Thomas wrote about communities and their moral codes of conduct, not as expressions of individual morality, but as sets of behavior norms that develop over time through repeated social interaction. Using insight into the power that social groups have on individuals, Thomas uncovered the fact that people respond not only to the *objective* features of a situation (what is real), but also to the *meaning* that the situation has for them. Furthermore, once we define situations in a certain way, our actions are often based on those definitions.

The meanings of situations are often shaped by social environments—including political structures, the economy, communities, social institutions like schools and the legal system, family, and peers. The discipline of criminology, which studies crime and the control of crime, is grounded in several disciplines, but primarily sociology. As a criminologist trained in the discipline of sociology, I believe that societies, subcultures, and social groups affect the individual offender, victim, and law enforcer, through definitions of social norms.

In the 1990s, I conducted research in Boston on two groups of people—police and gang-involved male juvenile offenders held at the Massachusetts Department of Youth Services (DYS). I was interested in how each group defined encounters with one another. My observations began when crime and policing were at a very critical stage in Boston. Juvenile homicides were at a record high, mostly due to gangs in inner-city neighborhoods that were competing for territory related to crack dealing. Police and juveniles were each frustrated by the relentless violence and record homicide rate, and they blamed one another for the situation. Within both groups were subcultures that reinforced myths and stereotypes. These myths were perceived as "real" and locked each group into routine, ritualized behaviors, guaranteeing that they would clash.

My goal was to uncover these patterns, expose untrue and misleading stereotypes, and use these new understandings to change behaviors on both sides. In related research involving a survey of 100 juvenile offenders (Guarino-Ghezzi & Kimball, 1996), my coauthor and I found that about two thirds were highly alienated from police, based on their responses to such questions as, "Would you go to the police if you believed your life was in danger?" I also found that the recidivism, or the rearrest, rate for youths who were highly alienated from police was 52%, compared with 28% for other youths—nearly twice as high.

One of my students at Stonehill College and I analyzed the data further, and we found that alienation from police was very strongly correlated with alienation from adults in general (Guarino-Ghezzi & Carr, 1996). We found that the youths who were the most alienated from police had the most frequent encounters with them. Negative encounters with police were at the center of their lives, ironically, because of the lives they chose. They could have avoided such unpleasant interactions by going to school or work, rather than hanging out in groups on the street. We found this to be a frequent contradiction, and came to realize that while the offenders claimed to hate police, they actually looked forward to negative confrontations as opportunities to reinforce peer bonds. When I suggested to DYS residents that I could arrange meetings with police in DYS facilities, they loved the idea.

At the same time, I learned that police painted ex–juvenile offenders with the same brush. If a juvenile offender returned to the community after a long program of rehabilitation, regardless of his or her willingness to reform, the individual was still a juvenile offender as far as the police were concerned. And if a juvenile offender became a victim of crime, the police showed less sympathy than if the juvenile were a law-abiding victim. When I asked what the Boston Police Academy was doing to prepare police recruits for encounters with juveniles on the street, I received a copy of the police academy's 26-week curriculum and was stunned to see only a few hours on juvenile crime, with no training on communication skills or adolescent development nor on the background of juvenile offenders. I realized that police were unaware how patterns of their behavior, such as mixed messages and overt stereotyping, were used by youths to justify increases in law-breaking, escalated youths' defiance toward police, and resulted in more arrests (Guarino-Ghezzi, 1994).

I contacted the deputy superintendent of the experimental neighborhood policing district in Boston (who, incidentally, had recently completed a degree in sociology at Boston College) and shared my concerns. He invited me to meet with two of his most progressive officers, who added to my observations. After several meetings, we all agreed that things needed to change.

Together, we established a program called Make Peace With Police (MPWP), which arranged communications sessions, role-plays, and other nonconfrontational encounters between police and gang-involved juveniles. As executive director, I oversaw 41 group meetings run by MPWP facilitators on a weekly basis. The groups ran from April 1995 until January 1997 and involved 70 youths and 35 police officers. The youths were part of ongoing programs in DYS, while the officers were assigned to attend the sessions as paid details. The officers usually rotated, with some electing to attend more than once. Pretest and posttest evaluation instruments were given to the youths and pretests to the officers, some of whom were then interviewed following the sessions. In addition, detailed notes of the group meetings were taken by students and later transcribed. These sessions helped us to understand the sources of hostility on both sides, but more importantly, they helped to create useful dialogue between juveniles and police.

In one session, a young gang member broke down in tears and said over and over, "The police have to squash the beef." This boy feared retaliation from other gang members but felt powerless to do anything about it. He believed that only the police could help him, but that the police didn't care enough about him to help. At the end of another session, a young man reminded two officers that they had met him before—in jail. They had stopped to talk to him while he was in a police lockup and gave heartfelt advice about how he was leading his life, and after our session he thanked them sincerely.

A key finding of the Make Peace With Police project was that without sincere efforts to establish relationships, juveniles lacking positive social bonds learn to define police negatively from their social environments. While some of the most defiant and hostile-seeming youths were biased against police based on their peer subcultures, our sessions became turning points for developing positive relationships. In one meeting, we brought in three officers, including a female named Officer Smith, to meet with about six kids. As usual, the participants were seated in a circle. The kids usually had their guard up, but this time, one of the kids outright refused to participate and he turned his chair to face outside of the circle. Officer Smith wasted no time moving her chair parallel to his. He sat there, arms folded, looking straight ahead. She began to smile and gently tease him, and started to poke him, saying that she was going to keep this up until she could get him to smile. After a few minutes he eventually smiled, then grinned, and then turned his chair back into the circle, to the amazement of everyone except for Officer Smith!

Similarly, police told us that they were surprised to learn that young gang members were really "just kids." An officer who arrived at a session feeling tough and somewhat angry at the youths quickly attached to a youth during the session who was visibly upset when mentioning a death in his family. The officer offered to help the youth find a job when he was released from the DYS facility. When asked what made him change, the officer explained that he had no idea how young and vulnerable the youths were because he'd never looked beyond the street-tough exteriors that were so common among groups of youths in high-crime communities.

Another officer told us about a DYS youth who went home on a weekend pass. The officer and the youth had met several times in Make Peace With Police sessions held at the youth's DYS program. The officer received a surprise that weekend when the youth recognized him in his cruiser and went out of his way to initiate a pleasant conversation. It was especially fulfilling because the boy was one of a small number who refused to even speak to police in his first Make Peace With Police session. The communication sessions provided a necessary bridge for redefining social norms. Both sides came to admit that negative police encounters with juvenile offenders can actually lead to *more* crime, not less.

We presented our work to scholarly audiences in the form of journal articles (e.g., Guarino-Ghezzi & Carr, 1996), and my colleagues and I incorporated the feedback given by our academic peers as we prepared materials for a police audience. The Boston Police Department published our booklet *Make Peace With Police: Myths and Rituals* (Godfrey, Guarino-Ghezzi, & Bankowski, 1997), which summarized our project findings and recommendations and became the basis for building effective communication skills in the Boston Police Academy. The booklet detailed how some of the long-term hostilities were rooted in simple miscommunications and how others were more complex.

In the end, I was deeply gratified because so many naysayers had warned us not to bring police and juvenile offenders together, predicting that the sessions would fail miserably or grow violent. I learned that police officers and juvenile offenders did not necessarily want to battle one another but felt pressured to do so by their social circumstances. Our sessions were able to initiate a change in social norms by altering those circumstances. They also reminded me, once again, that sociological tools can be used to make a positive impact on society.

References

Godfrey, K., Guarino-Ghezzi, S., & Bankowski, P. (1997). *Make peace with police: Myths and rituals*. Boston, MA: Boston Police Academy.

Guarino-Ghezzi, S. (1994). Reintegrative police surveillance of juvenile offenders: Forging an urban model. *Crime & Delinquency, 40*(2), 131–153.

Guarino-Ghezzi, S., & Carr, B. (1996). Juvenile offenders versus the police: A community dilemma. *Caribbean Journal of Criminology and Social Psychology, 1*(2), 24–43.

Guarino-Ghezzi, S., & Kimball, L. (1996, April). *Transitioning youth from secure treatment to the community*. Boston, MA: Department of Youth Services.

DISCUSSION QUESTIONS

1. Many of Michele Wakin's students had never interacted with homeless people before. How did their doing so in her class impact their perceptions of homeless Americans? Why do you think these interactions had such an influence on them?

2. Can you think of a group of Americans who have relatively little power (like homeless Americans) with whom you have never interacted? Why or why not? How do you think your interacting with such a group might influence your opinion of them? Would you be willing to "step outside your comfort zone" like some of Wakin's students did and participate in such an interaction? Why or why not?

3. Discuss the difference between service work and social activism that Shelley White describes. Which do you feel more comfortable carrying out? Why? How has *your* socialization process influenced this? Which has the potential to make more of an impact on society, service or activism? Why?

4. White describes the international service trip that she took as an undergraduate as "personally transformative." Why did it have such a major impact on her? Have you participated in a similarly transformative experience in high school or college? If so, what was it—and why was it transformative for you? How, if at all, did it challenge the way you view the world and your place in it? If you have not participated in a similar experience, why do you think that is?

Does it offer any insight into your own socialization experience that you have not yet participated in such a trip?

5. Were you at all surprised by the research findings Susan Guarino-Ghezzi discovered before she began her Make Peace With Police program? If yes, which ones—and why? If not, why not?

6. How does the Make Peace With Police program illustrate how social norms can be changed by altering social circumstances? Describe how changing social circumstances at your school might lead to a positive change in social norms (pick an issue you care about where you believe change is needed).

7. Each of the pieces in this chapter describes, in some way, the power of socialization. However, many Americans like to believe that they are not influenced by others. Pretend you are a Sociologist in Action who is confronted by people who do not believe that their behavior can be influenced by those with whom they interact. How might you use these articles to try to convince them of the influence of socialization?

8. Both Wakin's and White's pieces show how college experiences can impact a person's attitude toward social activism. Imagine you are advising a college president who believes it is the responsibility of higher education to help students become knowledgeable, effective participants in our democratic nation. How would you use these two articles to help make the case for both curricular and extracurricular efforts to promote social activism on college campuses?

RESOURCES

The following Web sites will help you to further explore the topics discussed in this chapter:

Introduction to Sociology/Socialization	http://en.wikibooks.org/wiki/Introduction_to_Sociology/Socialization
PBS: Nature vs. Nurture Revisited	http://www.pbs.org/wgbh/nova/body/nature-versus-nurture-revisited.html
Socialization and the Self, by Richard T. Schaeffer (2010)	https://globalsociology.pbworks.com/w/page/14711256/Socialization%20And%20The%20Self

To find more resources on the topics covered in this chapter, please go to the Sociologists in Action Web site at **www.sagepub.com/korgensia2e.**

Chapter 6

Deviant Behavior

Deviant behaviors, like norms and values, are socially constructed and vary over time and from society to society. While a few acts, like incest and cannibalism, are considered taboos and viewed as deviant (and abhorrent) in almost every society, some behaviors are considered "normal" in some societies but deviant in others. For example, while 50 years ago in the United States it was common to see pregnant women smoking, today an expectant mother who smokes would be labeled deviant. Likewise, today, relieving oneself in public is an accepted practice among men in some societies but seen as a deviant act in others. Finally, while some acts of deviance can damage the fabric of society and threaten its stability, other acts of deviance, if done in opposition to unjust or harmful norms, can promote needed social change.

In "Reducing Hate and Prejudice on Campus: A Sociologist's Contributions," Jack Levin describes his work to reduce hate crimes, one of the forms of deviance that threaten the stability of a society. In particular, he provides an overview of his work to curb hate crimes on college campuses. He offers evidence that such acts of deviance happen all too frequently. In Levin's words, while it is a good thing that campuses are becoming more diverse, "[I]t is not enough for the members of different racial or ethnic groups to have contact with one another. Everything depends on the quality of that contact." Levin has worked tirelessly to develop and promote programs that foster "deliberately created, structured opportunities for members of society to interact optimally on a cooperative and intimate basis with people who are different." These types of interactions encourage greater cross-ethnic/racial understanding and cooperation and create environments in which hate crimes are less likely to occur.

David S. Kirk's, "Using Evidence-Based Research to Inform Public Policy: Lessons from Hurricane Katrina," provides another powerful example of how sociological tools can guide social policies and programs. In this piece, Kirk relates how the devastation wrought on New Orleans, Louisiana, by Hurricane Katrina provided the makings of a natural experiment on the impact of residence on recidivism. His findings reveal that prisoners who did not return to their old neighborhoods upon release (because of the damage done to them by Hurricane Katrina) were much less likely to be back in prison one year after their release than those who did return to their old home environments. Kirk is now undertaking further research on this topic in order to "reveal to government stakeholders whether enacting large scale changes in the way criminal justice is practiced may . . . enhance public safety."

In "The Politics of Protest Policing," Alex S. Vitale discusses how certain police tactics against demonstrators can actually promote, rather than prevent, negative incidents during public protests. His work studying police responses to acts of deviance, public protests against the War in Iraq, and the 2004 Republican National Convention in New York City helps elucidate how different police tactics can promote *either* peaceful *or* violent protests. His work "generated extensive press coverage and is part of a now growing literature on how the police in the United States handle large demonstrations." Most importantly, it can now help law enforcement officials develop and use procedures that lower the likelihood of violence during protests.

In the last Sociologist in Action piece in this chapter, "From *Damaged Goods* to Empowered Patients," Adina Nack shares her story of how her own diagnosis with a cervical human papillomavirus (HPV) infection led her to recognize, firsthand, the stigma associated with women who have sexually transmitted infections (STIs) and the need to address this social issue. It also prompted her to take action by conducting a study of women with genital herpes or HPV that resulted in a number of articles and her book *Damaged Goods* (2008). Being willing to publicize her own HPV status has helped Nack enable others to see the connection between the personal troubles of women with STIs and the societal issues of sexism and the stigmatization of women. She states, "If 'knowledge is power,' then I hope that my research, writing projects, and applied collaborations empower not only STI patients but also increase the chances that those who struggle with stigmatizing illnesses can enjoy healthier and happier lives." Nack argues "that de-stigmatizing STIs, in all social venues, requires us to challenge traditional or sexist norms about sexual relationships and sexual health." In the process, she has given both hope and a voice to women with STIs.

REDUCING HATE AND PREJUDICE ON CAMPUS: A SOCIOLOGIST'S CONTRIBUTIONS

Jack Levin

Northeastern University, Boston, Massachusetts

Jack Levin, PhD, is the Brudnick Professor of Sociology and Criminology at Northeastern University in Boston, where he co-directs its Center on Violence and Conflict. He has authored or coauthored 30 books, including *The Violence of Hate: Confronting Racism, Anti-Semitism, and Other Forms of Bigotry* (3rd ed., 2010, with Jim Nolan). Levin recently received a major award from the American Sociological Association for his efforts to increase the public understanding of sociology. He has spoken to a wide variety of community, academic, and professional groups, including the White House Conference on Hate Crimes, the Department of Justice, the Department of Education, the Organization for Security and Co-operation in Europe's Office for Democratic Institutions and Human Rights (with a membership of 59 countries), the National Association of Hostage Negotiators, and the International Association of Chiefs of Police.

Our educational institutions are, for good reason, generally regarded as bastions of enlightenment and respect for diversity. Indeed, students on most campuses are exposed to a broad range of ideas, speakers, faculty members, and fellow students. They are encouraged to express dissenting points of view and to interact with diverse fellow students and faculty members. They might work or study in other countries, or do community service among diverse populations.

Yet our image of the college experience may also underestimate seriously the intergroup hostility and conflict that, on occasion, can come to define relationships on a campus. In a highly competitive environment, where students vie for grades, scholarships, jobs, organizational budgets, and popularity, they may regard one another not as allies but as opponents competing for scarce resources. In one recent survey of 10 campuses, it was determined that more than half of all students personally experienced or witnessed bias incidents—graffiti, verbal insults, physical threats, or physical assaults—targeting individuals because of their group identity (Campus Tolerance Foundation, 2009).

We shouldn't be shocked, then, when we read about hate crimes on a campus being directed against a student or a faculty member based on his or her race, religion, national origin, sexual orientation, gender, or disability status. Several examples of campus hate crimes and other acts of bias in just one recent academic year—2010—bring this point home. Swastikas were painted at the University of California, Davis, on the dormitory door of a Jewish student and on several campus buildings. The UC Davis campus center for lesbian and gay students was also vandalized (Rosenhall, 2010). A community college in Ohio "beef[ed] up security and provid[ed] alternative temporary living and sleeping arrangements for its black students after a note threatening to kill them on Feb. 2 appeared on a campus bathroom wall" (Sinclair, 2010). The editorial staff of the independent student paper of Notre Dame, *The Observer*, apologized after publishing a cartoon that implicitly condoned gay bashing (2010). We also know that, since the 9/11 attack on the United States, Jewish, Muslim, and Sikh students on college campuses around the country have been at greater risk for verbal and physical assaults (The Pluralism Project, 2005).

As a sociologist who researches these topics, I have been asked to speak on hate, hate violence, or hate crimes to students or faculty on numerous college campuses (and high schools) both in the United States and abroad. Hopefully, my presentations have inspired an initial reduction in hostility between groups, but there has to be more than a one-shot event to have any longer-term impact.

I always stress the incredible power of coalitions—temporary alliances of students who put aside their differences and come together to work toward the furtherance of their common goals. Special-interest groups on campus—the gay and lesbian alliance, Latino center, international student association, black student union, Vietnamese student alliance, and so on—are usually essential for providing underrepresented students with what they require in order to stay in school but cannot seem to get from the wider campus community. At the same time, however, there should also be curricular and extracurricular opportunities for diverse students to come together to cooperate in harmony and peace—this is the power of coalition building. At many colleges and universities, students from diverse backgrounds have organized rallies against violence, put on food and music festivals, and held speaker series that defend or celebrate *all* of their group memberships collectively.

I have urged schools to put interdependence into their curricular and extracurricular activities. It is not enough for the members of different racial or ethnic groups to have contact with one another. Everything depends on the quality of that contact. Most research into the impact of intergroup contact on prejudice has supported the notion that the good feeling that develops between cooperating friends from different groups actually generalizes in two ways. First, in many cases, it generalizes from the few immediate intimates to the entire group to which the few intimates belong. Second, individuals who come through contact to reduce their prejudice toward the members of a particular ethnic group seem to be more willing to interact with the members of other ethnic groups generally. In other words, intergroup contact seems to reduce not only negative attitudes and feelings toward the cooperating group, but also the general phenomenon known as *ethnocentrism,* in which an individual believes in the superiority of his or her own ethnic group and holds a generalized hostility toward the members of other ethnic groups.

In the Brudnick Center on Violence and Conflict at Northeastern University, we have developed a number of projects and programs for the purpose of preventing or responding to the presence of hate violence on campus. More than a decade ago, my colleague Will Holton and I teamed up to teach an experimental sociology course (Social Conflict and Community Service) that took teams of undergraduate students out of the traditional classroom to provide service—under optimal circumstances for the reduction of stereotyped thinking—to the local community. Our primary objective was to broaden students' perspectives, to give them an opportunity to interact with people of different races, ethnicities, or religions and to do so in a spirit of cooperation, civility, and goodwill. My colleague and I took pains to provide our students with a positive experience in the community—one that would not inadvertently reinforce their stereotypes and preconceptions. For example, if a student did service in a school, he or she would work in an honors class as well as a lower-track classroom or resource room.

Every week, each student in the course, as a member of a team, performed five hours of community service, and the group met together as a class for two hours to discuss related issues. In addition, students wrote journals summarizing their community service experiences for the week and a more inclusive term paper at the end of the course. Our final class meeting consisted of oral team presentations in which students summarized their community experiences and reflected on how those experiences had changed their own feelings and thinking about diversity.

Because they have grown up shielded from those who are different, many of the students in our course were familiar with people from other racial and cultural groups only as the stereotypes they saw on television or in movies. Their participation in community service learning provided an opportunity to interact cooperatively in a positive context with a wide range of individuals from other groups. Some of our students learned a good deal from being part of a project team whose members were diverse. At the same time, they were made aware of the existence of poverty and homelessness, flaws in the criminal justice system, prejudice and discrimination, and their own mortality. An unexpected advantage of our course for many of the students was that it taught them that they are not at the center of the universe. As one of our students concluded after spending 10 weeks working with Boston teenagers, "The greatest content of learning in this course was about myself. I was forced to explore my own prejudices and those of others like me." Social Conflict and Community Service has now become a permanent part of the curriculum and is taught annually in the Department of Sociology at Northeastern.

On almost every campus in the United States, it is possible to locate at least a few students who are willing to take responsibility for organizing rallies, demonstrations, festivals, or clubs in which diverse elements on campus are brought together in a spirit of cooperation. In their exceptional zeal, however, such students may feel alone and unappreciated.

At the Brudnick Center, I collaborated with Gordana Rabrenovic and with Steve Wessler's Center for the Prevention of Hate Violence to bring hundreds of such exceptional college students to Boston for the purpose of attending a National Student Symposium where they received awards for their efforts at combating hate and prejudice. Three hundred students representing more than 70 colleges and universities from more than 22 states plus the District of Columbia and the province of Quebec attended symposia in March 1999 and in April 2005. All had been nominated for their good work by the dean of students on their campuses. The symposia were funded by the U.S. Department of Education and the U.S. Department of Justice.

The agenda for each symposium included roundtable discussions, skill-building workshops, receptions, and an awards ceremony. Our objective was twofold: first, to recognize and reward college students who work to combat hate and, second, to let such students discover that they are not alone, that they have plenty of company and are appreciated by others.

Through my affiliation with the Brudnick Center, I continue to engage in programs and projects to reduce hate and prejudice. Students who are different (especially gay students) are frequently targeted for bullying in middle and high schools. In November 2009, I testified at the State House of the Commonwealth of Massachusetts in favor of a bill requiring all schools in the state to institute effective procedures against bullying. During the same hearing, two mothers of youngsters who had been bullied into committing suicide poignantly urged the legislature to take action. In March 2010, the Massachusetts Senate finally passed a version of the anti-bullying bill and prepared it to be signed into law.

More recently, I was asked to participate in the Teenage Interfaith Diversity Education Conference, to be held over Memorial Day weekend 2010 on our campus in collaboration with the Brudnick Center, but organized and attended by high-school students themselves. My role was to give a keynote presentation in which I addressed the value of diverse students cooperating toward the fulfillment of their shared objectives.

Recognizing the power of intergroup contact to bring diverse segments of the population together in peace and harmony, we simply cannot afford to leave such occasions to chance. Piecemeal efforts to create optimal contact experiences will result in trivial improvements in the overall social climate. Instead, we need more deliberately created, structured opportunities for members of society to interact optimally on a cooperative and intimate basis with people who are different. Hopefully, my own role in the process will make a worthwhile contribution to the overall effort.

References

Campus Tolerance Foundation. (2009). *Campus Tolerance Foundation.* Retrieved from http://www.campustolerance.com/campus_tolerance_media.pdf

The Observer. (2010, January 15). Responsibility for offensive comic [Staff editorial]. Retrieved from http://www.ndsmcobserver.com/viewpoint/responsibility-for-offensive-comic-1.1011631

The Pluralism Project. (2005). *Research report: Post 9/11 hate crime trends: Muslims, Sikhs, Hindus, and Jews in the U.S. (2005).* Retrieved from http://pluralism.org/reports/view/104

Rosenhall, L. (2010, March 15). More swastikas found on UC Davis campus. *Sacrobee.com.* March 15. Retrieved from http://www.sacbee.com/2010/03/03/2579662/more-swastikas-found-on-uc-davis.html

Sinclair, H. (2010, January 28). Hate crime has Ohio college on high alert. *Allheadlinenews.com.* Retrieved from http://www.allheadlinenews.com/articles/7017656731

USING EVIDENCE-BASED RESEARCH TO INFORM PUBLIC POLICY: LESSONS FROM HURRICANE KATRINA

David S. Kirk

University of Texas at Austin

David S. Kirk, PhD, is associate professor in the Department of Sociology and a faculty research associate of the Population Research Center at The University of Texas at Austin. His research on Hurricane Katrina and prisoner reentry was awarded the James F. Short Jr. Distinguished Article Award from the Crime, Law, and Deviance Section of the American Sociological Association. Kirk's recent research has appeared in *American Journal of Sociology, American Sociological Review,* and *Criminology.*

I have spent much of my professional career using the tools of sociology to examine the myriad consequences of criminal justice policies in the United States. The United States is one of the most punitive countries in the world, with an official incarceration rate that tops all countries reporting an official rate (Walmsley, 2009). One in every 100 adults in the United States is in prison or jail at this very moment, with more than 1.5 million individuals serving time in state and federal prisons and another 760,000 in local jails (Glaze, 2010; Pew Center on the States, 2008). The repercussions of mass incarceration become apparent when considering the fact that 730,000 prisoners are released each year from U.S. prisons (West, Sabol, & Greenman, 2010). Research suggests that up to half of releasees have been in prison before. In fact, by some estimates, two-thirds of returning prisoners are rearrested within three years of prison release, and more than half are reincarcerated (Langan & Levin, 2002). These staggering figures should not be separated from the social context in which prisoners return. Research reveals that ex-prisoners tend to be geographically concentrated in a relatively small number of neighborhoods within metropolitan areas, often returning to the very same neighborhoods where they got into trouble with the law in the past.

If criminal behavior is influenced by the types of neighborhoods we live in, then it would seem counterproductive to prisoner reintegration for ex-offenders to return to the same neighborhoods where they got into trouble with the law in the past. Upon exiting prison, ex-offenders who return to home neighborhoods often fall into the same habits and routines that got them into trouble in the first place. Too often, the only thing that has changed for many ex-prisoners is the ever-growing stigma of their criminal past. Thus, it is not surprising that large proportions of ex-prisoners end up back in prison within just three years.

These well known facts about crime and justice in the United States serve as the backdrop of my research on prisoner reentry, the process of leaving prison and returning to the community. I have attempted to use the tools of sociology—including a natural experiment, geographic information systems, a substantive knowledge of life-course criminology, and an awareness of the importance of social context—to examine the efficacy of alternative ways of reducing crime and recidivism (i.e., the relapse into some form of criminal behavior).

A typical experiment is a study in which an intervention is intentionally introduced in order to observe the effect of the intervention on an outcome (Shadish, Cook, & Campbell, 2002). An example from medical research is giving a treatment group an experimental drug and a control group a placebo, and then observing differences in subsequent health across groups. A natural experiment is based on a naturally occurring (i.e., unintentional) event that induces some kind of intervention. The tragedy of Hurricane Katrina, which devastated the gulf coasts of Louisiana and Mississippi in August 2005, afforded me a unique opportunity to examine what would happen if ex-prisoners did not return home to their old neighborhoods upon exiting prison as they typically do. Katrina provided a natural experiment for investigating the importance of residential change because it forced some people to move who otherwise would not have.

The idea for the study arose from exploring the devastation to New Orleans from Hurricane Katrina. I have extended family in New Orleans, many of whom evacuated to the Washington, DC, area where I was living at the time. My family returned to New Orleans around Thanksgiving of 2005, roughly three months after the hurricane struck. My first visit to New Orleans to see my family following Hurricane Katrina was in December 2005. During that visit and subsequent visits, out of both morbid curiosity and my interest in exploring social contexts, I would drive around the city. The devastation to entire neighborhoods was shocking. I could not even fathom the extent of disruption to the lives of tens of thousands of individuals.

While no demographic group was unaffected by Hurricane Katrina, it occurred to me while I drove around the streets of New Orleans that some of those neighborhoods hardest hit by property destruction were the very neighborhoods where ex-prisoners typically resided upon release from prison. My ability to make this insight was surely made possible by my training in urban sociology, and the understanding that social problems tend to tightly cluster in geographic space. Many individuals coming out of prison soon after Katrina would not be able to return to their old neighborhoods. Yet, moving is not necessarily a bad thing. A well-known argument from the literature on life-course criminology is that changes in behavior are often induced by "turning points" that provide a fresh start in an individual's life. A turning point can be an opportunity or life experience that redirects a previous behavioral pattern, such as criminal offending. Typical examples of turning points include marriage, birth of a child, and enlistment in the military. It occurred to me while driving around New Orleans that residential change may also be a turning point in ex-offenders' lives. Residential change may serve as a catalyst for sustained behavioral change by providing an opportunity for individuals to separate from the former contexts, situations, and criminal associates that facilitated their prior criminal behavior.

To investigate these ideas, I knew that I needed to collect data on the residential patterns of prisoners released from Louisiana prisons both before and after Hurricane Katrina (for comparative purposes) as well as data on recidivism among these individuals. In this kind of research, data can be tough to come by. In my case, I was fortunate to connect with justice administrators who provided access to their data. They, too, were interested in understanding the repercussions of residential change on rates of criminal behavior.

One of the first steps in my research process was to use geographic information systems to map the street addresses where ex-prisoners resided immediately upon exiting prison. In comparing the maps for the pre-Katrina and post-Katrina time periods, both visually and statistically, I noticed that there had been a substantial shift in the patterns of residence following Hurricane Katrina. As I had hypothesized, prisoners exiting incarceration following Hurricane Katrina were much less likely to reside in the New Orleans' neighborhoods where they resided prior to incarceration.

With evidence on changing patterns of residence, I next used statistical analyses to assess the repercussions of residential change. I found that an estimated 26% of male offenders who returned to the same parish (parish is the equivalent of a county) where they resided prior to incarceration were reincarcerated within one year of release from prison (Kirk, 2009). By comparison, only 11% of male offenders who moved to a new parish faced reincarceration one year after leaving prison. Based on these results, I conclude that separating individuals from their former residential environment reduces their likelihood of recidivism. Moving allows an individual to separate from the peers and routine activities that contributed to his or her criminal behavior in the past.

While results from this natural experiment provide some initial evidence on the importance of residential change, in the interest of good science I have been engaged in several research projects over the past few years designed to validate these initial findings. My results have provided further support for my initial conclusions (Kirk, 2011, 2012).

Armed with mounting evidence about the dire consequences if ex-prisoners return home to former neighborhoods, I have recently spent time engaging in what the National Institute of Justice calls *translational criminology*. The idea is to translate evidence-based scientific discoveries into policy and practice. For this purpose, I have shared my discoveries with a variety of audiences, including the National Institute of Justice, the Prisoner Reentry Institute, the Stockholm Criminology Symposium, and the Austin/Travis County Reentry Roundtable. More informally, I have communicated my ideas and results through ad-hoc meetings to criminal justice practitioners and policy makers in several states.

One critical component of disseminating information about scientific discoveries is to communicate the implications of the research. For instance, in most states, prisoners released on parole are legally required to return to their county of last residence, thus contributing to a return to old neighborhoods. Thus, parole policies, while designed to enhance public safety, may in fact undermine it. Given evidence that residential change fosters the path to behavioral change, one implication of my research that I have discussed with key stakeholders is that removing the institutional barriers to residential change may enhance public safety in aggregate by lowering recidivism. Additionally, providing incentives for individuals to move to new neighborhoods, such as public housing vouchers, may also benefit public safety.

The subject of crime is a politically charged issue that stokes much passion in the general public. Indeed, in a punitive country like the United States, it is not politically popular to be soft on crime or terrorism. In reality, despite good science demonstrating the efficacy of policy changes, modifications to existing social policies are not usually immediate. It is challenging to break the status quo. Nevertheless, I have been engaged in discussions with several state prison systems to implement demonstration projects with several hundred parolees to test whether allowing and incentivizing residential moves among ex-prisoners will reduce their likelihood of recidivism. These demonstration projects will show whether my findings from a unique natural experiment (i.e., Katrina) are applicable to a real-world policy environment. These demonstration projects will also reveal to government stakeholders whether enacting large scale changes in the way criminal justice is practiced may in fact enhance public safety. Thus, in my experience, redesigning public policies is part of a methodical process that involves good science, communication of results, and further testing in a real-world environment. The outcome of this process is fundamental to creating a just society.

I was drawn to the discipline of sociology out of curiosity and a sense of social justice. I was curious how the social world operates, and chose a profession in which I could answer for myself and others foundational questions, such as "What is the purpose of society?" but also more practical questions like, "Is there a better way to undertake parole in the United States?" To me, if we can understand the causes and consequences of social problems, then maybe we can do something about eliminating them. Sociology has provided me the tools necessary for discovering effective solutions to society's problems.

References

Glaze, L. (2010). *Correctional populations in the United States, 2009*. NCJ 231681. Washington, DC: U.S. Department of Justice, Bureau of Justice Statistics.

Kirk, D. S. (2009). A natural experiment on residential change and recidivism: Lessons from Hurricane Katrina. *American Sociological Review 74*(3), 484–505.

Kirk, D. S. (2011). Causal inference via natural experiments and instrumental variables: The effect of 'knifing off' from the past. In John M. MacDonald (Ed.), *Measuring crime and criminality: Advances in criminological theory*, (Vol. 18, pp. 245–266). New Brunswick, NJ: Transaction Publishers.

Kirk, D. S. (2012). Residential change as a turning point in the life course of crime: Desistance or temporary cessation? *Criminology.* Forthcoming.

Langan, P. A., & Levin, D. J. (2002). *Recidivism of prisoners released in 1994.* NCJ 193427. Washington, DC: U.S. Department of Justice, Bureau of Justice Statistics.

Pew Center on the States. (2008). *One in 100: Behind bars in America 2008.* Washington, DC: The Pew Charitable Trusts.

Shadish, W. R., Cook, T. D., & Campbell, D. T. (2002). *Experiment and quasi-experimental designs for generalized causal inference.* Boston, MA: Houghton Mifflin.

Walmsley, R. (2009). *World prison population list* (8th ed.). London, UK: International Centre for Prison Studies. Available at http://www.prisonstudies.org/info/downloads/wppl-8th_41.pdf

West, H. C., Sabol, W. J., & Greenman, S. J. (2010). *Prisoners in 2009.* NCJ 231675. Washington, DC: U.S. Department of Justice, Bureau of Justice Statistics.

THE POLITICS OF PROTEST POLICING

Alex S. Vitale

Brooklyn College, Brooklyn, New York

Alex S. Vitale is associate professor of sociology at Brooklyn College. From 1990 to 1993, he was a staff analyst at the San Francisco Coalition on Homelessness before moving to New York City, where he received his PhD from the City University of New York Graduate Center. He authored *City of Disorder: How the Quality of Life Campaign Transformed New York Politics* (2009) and has published in *Urban Affairs, Current Sociology, Policing and Society, Mobilization,* and *Criminology and Public Policy.* He is regularly quoted as an expert on protest policing in the media including the *New York Times, Wall Street Journal, Atlantic Magazine, Salon, The Financial Times,* and NPR's *Talk of the Nation.* In 2010, he had a Fulbright scholarship to study police innovation in response to economic and political liberalization in Seoul, South Korea.

On February 15, 2003, millions of people around the world took part in the largest coordinated day of action on record to oppose a possible war in Iraq. Protests with close to or over one million people took place in London, Madrid, Rome, Berlin, and Paris. Hundreds of others occurred in cities in the United States and around the world, including New York, Los Angeles, Chicago, and San Francisco.[1] The demonstration in New York had between 100,000 and 400,000 participants.[2] I was one of those participants, who got up early that morning and went with some friends to one of the staging areas where people gathered before marching in groups to the rally area on First Avenue along Manhattan's Upper East Side.

As with many demonstrators, I was concerned about how the police would handle such a large demonstration, especially since they had refused to grant permits for people to march, creating the possibility of confrontations as people attempted to get to the rally area. Unfortunately, that's exactly what happened. As people approached First Avenue, they were denied access and forced to walk many blocks north to enter the rally area from constantly changing entry points. As crowds grew larger, many people moved into the surrounding streets while trying to make their way north. This in turn prompted the police to blockade streets and sidewalks in an effort to disperse what they viewed as spontaneous unpermitted marches. The result was a series of confrontations with the police in which nonviolent demonstrators were attacked with police horses, pepper spray, and baton-wielding police officers. After it was all over, hundreds had been arrested, many were injured, and I and thousands of other people had been denied our right to demonstrate against the impending war in Iraq.

Most of my sociological research up to this point had been about urban politics and the role of the police in handling the homeless, "squeegee men (and women)," and other disorderly individuals and groups. I was interested in studying the rise of "broken windows" policing, which is based on the idea that the police need to put more effort into controlling disorderly behavior, such as aggressive panhandling, sleeping in parks, and drinking in public. By doing so, the hope is that it would create a sense that streets were safer and less inviting to more serious types of crime. According to this philosophy of policing, the police needed to take a strong zero-tolerance stand against even minor violations of the law in order to prevent more serious law-breaking from occurring.

---

[1]Steffan Walgrave and Joris Verhulst, "Government Stance and Internal Diversity of Protest: A Comparative Study of Protest Against the War in Iraq in Eight Countries" (*Social Forces*, 87, 2009), 1355–1387.

[2]Chris Dunn et al., *Arresting Protest: A Special Report of the New York Civil Liberties Union* (New York Civil Liberties Union, 2003).

As I was watching the policing of the protest in front of me, it occurred to me that I was seeing the broken windows philosophy being applied to the policing of demonstrations. The police were taking extreme measures against essentially peaceful crowds because of very minor violations of the law. It seemed clear to me that the police were taking this zero-tolerance approach because they believed that the crowd might engage in more serious forms of illegal or disruptive behavior if they didn't aggressively control their movements and actions from the beginning.

When I got home, I decided to write up a short description of this insight. I explained a little bit about the broken windows theory of policing and then showed how it was being used to police the demonstration and how that had, in turn, led to the major confrontations that occurred that day. I then sent it around to friends, colleagues, and people I knew who had attended the demonstration. I also posted it on a couple of listservs and Web sites related to the demonstration. Within a few days, Donna Lieberman, the executive director of the New York office of the American Civil Liberties Union (NYCLU), called me. She said she had received several copies of my analysis by e-mail and had also been handed several printed-out copies from staff in her office. In addition, they had received dozens of e-mails, phone calls, and letters from their members complaining about how the police had treated them that day. As a result, she had decided that the NYCLU should write a report about what happened, and she invited me to work with them on it.

As we gathered information for the report, we solicited others to write about their experiences at the demonstration. In all, we reviewed 335 written accounts, along with press coverage and videotapes. In the end, we produced a report that detailed negotiations between the police and protest organizers before the event and what happened at the demonstration, and compared it to other protests around the world. We found that the police used high levels of force against nonviolent demonstrators, whose only violation of the law was that of walking in the street without a permit. We also discovered that in no other large city in the world had police responded to Iraq war protestors with aggressive and restrictive techniques like those of the New York Police Department (NYPD). We distributed thousands of copies of the report and received extensive press coverage. We also learned that the NYPD distributed copies of the report to many of its commanding officers.

Just before our report came out, New York City mayor Michael Bloomberg announced that the city would be hosting the 2004 Republican National Convention (RNC), which meant that the city could expect to see numerous large demonstrations and the possibility of a repeat of some of the problems we had documented. In response, the NYCLU asked me to join them in pursuing three strategies to try to prevent this.

First, the NYCLU undertook a lawsuit against the police in federal court, asking that some of the restrictive measures the NYPD used to control demonstrators be declared unconstitutional. During the course of the federal lawsuit, I was asked to testify about the nature of some of the police tactics used and their effect on demonstrators. In the end, the court ruled that the police had to change the way they used barricades to pen in crowds and restrict access to legal demonstration areas.

Second, the NYCLU went to the city council to ask them to put pressure on the mayor and the NYPD to respect the rights of demonstrators. I prepared and delivered written testimony to city council committees about the problems I saw with how the police handled the February 15 demonstration, about the problems I anticipated might arise during the RNC, and about how to correct them. This overall effort generated additional media coverage and resulted in the passage of a resolution by the city council calling on the NYPD to respect the constitutional right of people to demonstrate at the RNC.

Finally, in the weeks leading up to the RNC, I worked with the NYCLU to begin training volunteers to systematically monitor the policing of all the demonstrations during the convention. This was the first large-scale project of this type. It combined the legal issues involved in the first amendment's protection of the right to assemble with sociological insights about the way the police and crowds operate. My role involved developing and implementing the training curriculum and supervising monitors in the field. In the end, I and the NYCLU staff trained over 150 people who were assigned to teams of two to four monitors to observe over 40 demonstrations across the city in a seven-day period.

During the RNC, over 1,800 people were arrested on a variety of minor charges. Some people were held for days in makeshift holding pens in a former bus depot without access to lawyers or a judge. We found that many of these people had been arrested "preemptively" by the police, meaning that they had not actually committed any illegal acts at the time of arrest. Instead, the police arrested them because they thought these people *might* be preparing to engage in illegal civil disobedience types of activity. Our documentation, along with that of others, resulted in many charges being immediately dismissed by the district attorney. In the end, over 90% of cases resulted in dropped charges or acquittals.

Fortunately, we did not see the police use the same levels of force against RNC demonstrators that we saw at the 2003 antiwar protest, and many of the demonstrations happened without incident. This suggested to me that there were two different approaches at work during the RNC. The first was a restrictive but mostly nonviolent command-and-control style, similar in many ways to what happened in 2003. Two differences from 2003, however, were that the size of crowds was not as large and more police were assigned, allowing them to micromanage demonstrations without having to fall back on the use of force. This change in approach may have been a result of the findings in our 2003 report. The second approach was a much more repressive "Miami model," which relied on preemptive arrests, long detentions, special weaponry, and surveillance of demonstrators. Our final report, entitled "Rights and Wrongs at the RNC," highlighted these two different approaches and raised serious concerns about the decision to arrest so many people who had not committed a crime and the treatment they received while in custody.

This report generated extensive press coverage and is part of a now growing literature on how the police in the United States handle large demonstrations, a field that barely existed when I wrote my first analysis in February 2003. According to Donna Lieberman of the NYCLU, this process of collaboration "has shown that criminologists have something to teach lawyers about the first amendment" (personal communication, October 26, 2004). For me personally, it has opened up new intellectual opportunities and given me a greater appreciation of the potential for action research. Since writing these reports, I've been invited to join an international collaboration studying large protests in the United States and Europe, which will allow us to perform much more sophisticated cross-national comparisons of protest policing than I was able to in 2003. Overall, I've come to think of myself as more of a "public sociologist," whose primary audience isn't limited to students in the classroom or the readers of professional journals, but includes policy decision makers and even the general public. In the same way, I try to train my students to think not just as social scientists, but as members of an informed public who have a responsibility to use knowledge to create a better, more just society.

FROM *DAMAGED GOODS* TO EMPOWERED PATIENTS

Adina Nack

California Lutheran University, Thousand Oaks

Adina Nack, PhD, has been active in sexual health education and research since 1994. Author of the book *Damaged Goods? Women Living With Incurable Sexually Transmitted Diseases* (2008), Nack has won awards for her research, teaching, activism, and public policy work. Currently she is a professor of sociology at California Lutheran University where she also serves as director of the Center for Equality and Justice. Nack lives in Ventura County with her daughter and her husband, José Marichal, to whom she is indebted for his support in working to publicly destigmatize sexually transmitted infections. Visit her online at www.adinanack.com.

As a 20-year-old, being diagnosed with a cervical human papilloma-virus (HPV) infection did not, at first, seem like a positive "turning-point moment" (Strauss, 1959) in my life. I had no idea that my illness experiences would inspire me to pursue a sexual health education career and ultimately become the foundation for my first sociological study. Back then, I thought that this virus heralded the end of my sex life and maybe marked the end of my fertility. Perhaps most jarring was that this illness made it hard to see myself as a "good girl" . . . someone who could some-day become a "good" wife and a "good" mother.

Shuffling out of the procedure room after receiving cryosurgery[3] from my gynecologist, I felt like *damaged goods*—not just physically, but psycho-logically and socially. Like most Americans, I had been socialized to believe that women who contracted sexually transmitted infections (STIs) were sluts: dirty, promiscuous, irresponsible, stupid sluts. Depressed and consid-ering a lifetime of celibacy, I continued my undergraduate education and found myself gravitating toward women's studies and sociology courses.

Feminists often say, "The personal *is* political,"[4] which pairs nicely with C. Wright Mills' (1959) assertion that the *sociological imagination* allows us to view our "personal troubles" within social contexts to reframe them as "public issues." Thanks to my supportive parents and their class privilege—as a family who could afford the best medical treatments—I could put my HPV concerns behind me before beginning graduate school, but my sociological concerns about sexual health policies and practices lingered.

As a PhD student, I volunteered as a sexual health peer educator and even-tually directed my university's sexual health education program. Presenting on STIs/HIV to audiences—from junior high students to college students—I met others who were STI-infected and not finding the emotional support they needed. After trying to start a support group (to which only one woman showed up), I realized there was something *sociological* going on: Support groups for other stigmatizing conditions were flourishing—from HIV-positive groups to 12-step programs for a variety of addictions. I conducted a survey and found that most female patients feared disclosing their STI diagnosis, even to other similarly infected women in a confidential setting.[5]

---

[3]The medical application of liquid nitrogen to freeze/kill HPV-infected cells.

[4]This statement is often traced back to Carol Hanisch's essay, "The Personal Is Political," origi-nally published in the Redstockings collection *Feminist Revolution* (New York, NY: Random House, 1979), 204–205.

[5]See pp. 98–100 in my 2000 article, "Damaged Goods: Women Managing the Stigma of STDs" (*Deviant Behavior*, 21: 95–121); also pp. 17–19 and Appendix B in my book *Damaged Goods? Women Living With Incurable Sexually Transmitted Diseases* (Temple University, 2008) for more about how this survey informed my research design.

I've always liked the idea of being a public sociologist: translating research-based findings that have applied value to nonacademic audiences who can use the new knowledge. So, for my dissertation, I identified a real-world problem I thought could benefit from evidence-based and theory-driven research. To get to the social-psychological "heart of the matter," I utilized symbolic interactionism and feminist theories to analyze in-depth interview data. I drew on feminist scholarship about gender norms of sexual behaviors and used symbolic interactionism as a lens through which to focus on how individuals intersubjectively formed meanings about STIs during social interactions with medical practitioners and with significant others.

I interviewed adults with medical diagnoses of genital herpes and/or HPV (human papillomavirus) infections. Trained as an ethnographic researcher, I found that one-on-one, in-depth, semi-structured interviews allowed me to provide participants with confidentiality when sharing sensitive illness narratives. But I'm not sure that all of the methodological training in the world could have allowed me to connect with my participants—to gain their trust and create rapport—if my own HPV infection had not provided me with complete membership status in this setting.

I initially focused on writing up my findings to present at academic conferences, with the goal of publishing in academic journals. Most academic journals have relatively small audiences of readers, so I was grateful when my first and second articles were reprinted in undergraduate readers for courses like introduction to sociology, deviance, and sexuality. I began to receive e-mails from students who identified with my findings. For example, Anne,[6] a senior at the University of Florida, e-mailed me the following:

> I have just gone through the most emotional/traumatic three years of my life and the title of your article ["Damaged Goods: Women Managing the Stigma of STDs"] is exactly what I have gone through. . . . I am really thankful for the work that you are doing in this field.

She and others told me that they felt like I understood what they'd been going through and wanted to know more about my findings.

A six-stage process of "sexual-self transformation" emerged from my analysis. The first five stages represented a series of problems caused by myths, misinformation, harmful interactions (with medical professionals

---

[6]Pseudonyms are used throughout this piece to protect the confidentiality of those who have written to me.

and significant others), and treatments (that were not always effective, often painful, and sometimes quite expensive). The final destination, the sixth stage of *reintegration,* represented an elusive but important goal—a new sexual self that was healthier and happier as a result of balancing risk-awareness with desires for intimacy.

Feedback like Anne's inspired me to write up the study as a book that would be accessible to the typical undergraduate student. With this goal, I saw my book[7] as a form of advocacy. Those living with STIs have recommended it to each other on sexual health discussion forums (e.g., the American Social Health Association[8]) and have reviewed it on STI-specific Web sites. One HPV blogger wrote, "I was expecting *Damaged Goods* to be something 'over my head,'" and then went on to say she had found the book to be "a new and enlightening reading, compelling." Nonacademic readers got it—sociological research was helping them understand the social-psychological impacts of being diagnosed with medically incurable (though treatable), highly stigmatizing diseases.

By applying feminist theories, I could explain why female STI patients suffered more than their male counterparts. In my articles and book, I argue that destigmatizing STIs requires us to challenge traditional and sexist norms about sexual relationships and sexual health. College students have given me feedback that they understand my sociological explanations of STIs as "personal troubles" and were also inspired to think about STI stigma as a "public issue":

> I was amazed at how insightful and helpful this book is to not only someone living with HPV or HSV, but also to all of those people who they encounter and are possibly affected by it (significant others, doctors, parents, friends, etc.). . . . The stigma that goes along with this situation is wrong, hurtful, and unfair. Nack's efforts to destigmatize this problem are impressive and encouraging and her words really have something for everyone to benefit from. The jokes need to stop. The ignorant comments need to stop. The stigma needs to stop.[9]

Sociological training helped me to see, name, and examine the dangers of mixing morality with medicine—these messages were starting to resonate among undergraduate and graduate students.

---

[7]Adina Nack, *Damaged Goods? Women Living With Incurable Sexually Transmitted Diseases* (Philadelphia, PA: Temple University Press, 2008).

[8]http://www.ashastd.org/phpBB/viewtopic.php?f=4&t=7017

[9]http://www.amazon.com/review/RYSNFBWBUDT1U/ref=cm_cr_rdp_perm

In order to get the word out beyond college classrooms, I had to turn my research into sound bites that could be used by journalists. Translating findings for mainstream media was key to becoming a public sociologist. Writing for nonacademic blogs and participating in a magazine writers' workshop helped me to develop these skills, and I sought out opportunities to be featured as a sexual health expert—on TV and radio, and in newspapers and magazines that were reaching local, national, and international audiences. I'm a grassroots activist who believes in the power of interactions. I had to be willing and able to make my research—and my self—accessible to a range of nonacademic audiences.

I publicly disclosed my own STI status in the methods sections of my first academic article because it seemed methodologically important for readers to have a sense of my potential biases. And, for the book, I included my own STI autoethnography as an appendix. TV producers, radio hosts, and reporters were drawn to my research, in part because I was willing to talk openly and my sexual partner was (1) uninfected and (2) willing to talk about our sex life on TV.[10] We've been willing to be "poster children" for HPV because we understand that personal narratives can entice viewers and readers into making sociological explorations of sensitive and controversial topics.

As a medical sociologist, I've worked to promote the individual-level and public health benefits of destigmatizing STIs. Research collaborations and service-learning projects with local organizations—like the HIV/AIDS Coalition of Ventura County, CA, and Planned Parenthood of Santa Barbara, Ventura, and San Luis Obispo Counties, Inc.—have given me opportunities to work with practitioners to improve sexual health care policies and practices and to advocate for comprehensive sex education.

Embracing a public sociology perspective motivated me to both produce and disseminate knowledge about sexual and reproductive health. I will keep working toward the goal of destigmatizing STIs because it represents not only the ideologically correct position but the position that we must embrace to improve individual-level and public health. If "knowledge is power," then I hope that my research, writing projects, and applied collaborations empower STI patients as well as increase the chances that those who struggle with stigmatizing illnesses can enjoy healthier and happier lives.

---

[10]First interviewed for a 1999 MTV episode of *Sex in the 90s* and more recently in a fall 2008 episode of the CBS daytime talk show *The Doctors*. Clip available on YouTube at http://www.youtube.com/watch?v=su7Hcdt3Irs.

DISCUSSION QUESTIONS

1. Jack Levin points out in his Sociologist in Action piece that "in one recent survey of 10 campuses, it was determined that more than half of all students personally experienced or witnessed bias incidents." Are you surprised by this finding? Why or why not?

2. What does Levin mean when he says, "It is not enough for the members of different racial or ethnic groups to have contact with one another. Everything depends on the quality of that contact"? Are there opportunities for high-quality interracial/ethnic contact on your campus? Why or why not? What can you do to help create more such opportunities?

3. How did David Kirk's sociological background enable him to recognize the makings of an excellent (though clearly angering) natural experiment while he was driving around New Orleans in December of 2005? Have you been in a circumstance where you think your sociological eye would have helped you to identify a natural experiment? If so, explain. If not, do you think you would have been able to use your sociological eye in the way that Kirk did when he made his observations?

4. According to Kirk, why is it "challenging to break the status quo," in terms of how our criminal justice system operates? If you were an elected official who had some influence over policies related to Kirk's work on recidivism, would you advocate for programs that would enable or encourage prisoners to move away from their home neighborhoods when they are released from prison? Why or why not? What might be the advantages and/or the disadvantages to your political career from your decision?

5. Have you ever participated in a protest against a norm you considered immoral? Why or why not? How do you think such protesters should be treated by law enforcement personnel? Why?

6. According to Vitale, how can the broken windows approach to dealing with protestors lead to major confrontations between protestors and police? Why is it to the benefit of both police and protestors to make Vitale's findings widely known and to make use of them?

7. When you were reading Adina Nack's piece, how did you react when you learned that she is living with an STI? What does this tell you, if anything, about your own socially constructed stigmas around STIs?

8. If you were living with an STI, would you have the courage that Adina Nack has had to be public about it and to speak out to educate others? Why or why not? What would be the danger of not speaking out?

9. How do Nack's and Kirk's pieces illustrate that it is difficult to "break the status quo"? What role do sociologists have in challenging social constructions and helping to create new understandings about stigmatized and under-represented groups? How have Nack and Kirk each done so?

10. How do Levin's and Vitale's pieces show how different forms of social interaction can help promote *or* alleviate violent deviant behavior? Before you read their essays, had you realized that people's behavior could be so dramatically influenced by their social environment? Why or why not?

RESOURCES

The following Web sites will help you to further explore the topics discussed in this chapter:

Crime and Social Deviance	http://www.sociosite.net/topics/right. php#CRIMI
Deviance Flash Cards	http://media.pfeiffer.edu/lridener/dss/ crimedev.html
Deviance Links	http://media.pfeiffer.edu/lridener/dss/ crimedev.html
Slideshare Deviance	http://www.slideshare.net/rcragun/ deviance-2996170
Sociological Theories to Explain Deviance	http://ww2.valdosta.edu/~klowney/ devtheories.htm

To find more resources on the topics covered in this chapter, please go to the Sociologists in Action Web site at **www.sagepub.com/korgensia2e.**

Chapter 7

Social Movements

Social movements are organized efforts by a significant number of people to promote social change. These efforts focus on shifting norms and values and the social and legal structures associated with them, which are considered unjust (e.g., the civil rights movement, the women's rights movement, etc.), or preserving norms and values deemed endangered (e.g., the segregationists, the Luddites, etc.). In this chapter, four Sociologists in Action enthusiastically describe their use of sociological tools in social movement efforts. As you will see, each used his or her sociological eye to uncover patterns of injustice and then participated in organized efforts to address those injustices.

Our first Sociologist in Action in this chapter, Ellen Kennedy, explains how her trip to Rwanda motivated her to "take a stand and do something" about genocide. In *"Never Again*, Must Mean *Never,"* she describes the anti-genocide social movement and her participation in it through her founding of and work with World Without Genocide. In the process, she illustrates how almost all social movements begin with a precipitating event and follow the stages of emergence, coalescence, bureaucratization, and decline (through failure or because they have succeeded and are no longer needed). Now in the bureaucratization stage, World Without Genocide works to protect innocent people, prevent discrimination, prosecute perpetrators, and remember those impacted by violence. She invites all of us to join her efforts to create a world without genocide.

Mikaila Mariel Lemonik Arthur writes about her efforts to assist campus-based movements to promote curricular change in "Change the World—Start at School." She points out that "before the 1960s, people simply did not do research on or teach about issues related to women, Asian Americans, or queer/LGBT people. The emergence of these new disciplines has led to broad changes in American academic and social life." Her research focuses on why some curricular change campaigns succeeded while others failed. She found that, "unlike in the case of movements targeting the political sphere, movements targeting colleges and universities find that their strategies must be *less* assertive [in] less favorable" campus environments. She has made her findings available for those working for change "as part of a toolkit of ideas and techniques they can use in their own campaigns" and asks, "How can *you* make your own institution a better place?"

Rob Benford describes in "A Campus Gun-Free Zone Movement" how he and a colleague at the University of Nebraska-Lincoln (UNL) created a successful movement to ban guns on their campus. In 1994, a survey of his Principles of Sociology students revealed that 14.2% of the male students and 4.6% of the female students had carried a gun or knife to campus. Realizing that a sizeable number of students sitting before him during each class period were likely armed made him think "that playing poker in a Wild West saloon might have been a safer vocation." Utilizing these findings and their sociological knowledge of social movements and complex organizations, Benford and his colleague "mobilize[d] campus leaders, including not only UNL administrators, but representatives of various constituency groups on campus," to bring about a "gun-free"—and much safer—campus.

Finally, Charles Derber vividly illustrates the career of an exemplary Sociologist in Action in "Social Movements and Activist Sociology." From his days in the civil rights movement being "taken hostage by some White racists, shot at on a country road when driving with Black activists, and slapped in jail by some White sheriffs with no love for Yankee boys," to organizing a National Student Strike Center against the Vietnam War, to engaging in the Battle of Seattle and anti-sweatshop activism, to his leadership in the Occupy Wall Street movement, Derber has spent his entire life using sociological tools to build social movements. The author of many captivating books on economic and political injustice (and how to effectively fight against them), Derber maintains that his best writing "has been animated and informed by my participation in real world struggles to change socially dominant and violent institutions." This piece is an example of Derber's prowess both as a writer and as a change-maker.

NEVER AGAIN MUST MEAN *NEVER*

Ellen J. Kennedy, Executive Director

World Without Genocide

Ellen J. Kennedy, Ph.D., is the founder and executive director of World Without Genocide at William Mitchell College of Law, St. Paul, Minnesota, committed to protecting innocent people; preventing genocide by combating racism and prejudice; advocating for the prosecution of perpetrators; and remembering those whose lives and cultures have been destroyed by violence. Kennedy's awards include Outstanding Citizen, 2009, from the Anne Frank Center, Jane Addams Outstanding Service, 2010, from the Midwest Sociological Society, and two awards for community service, 2010, University of Minnesota.

I went to Rwanda in 2005. I had taught about the Rwandan genocide for several years and I wanted to understand how nearly a million people were killed in a hundred days while the world stood by.

A young Rwandan woman, Alice Musabende, traveled with our group as our guide and interpreter. One night, sitting under a starlit sky, I asked what happened to her family in 1994 during the genocide. She said she was 14 years old in 1994. One ordinary spring morning, her mother sent her on an errand to her cousin's house in the next village and told her to come right home. Alice went to her cousin's, but she stayed overnight. When she got home, she found that her grandparents, parents, 12-year-old sister, and 9-year-old and 2-year-old brothers had all been killed. Alice became an orphan of genocide.

Alice's story struck my heart—because she and my daughter Louisa are the same age. That spring of 1994, when Alice's world collapsed, my daughter's life was filled with the warmth of a loving family.

Soon after Alice told me that story, she and I visited a memorial in Rwanda to those who had perished. After I saw the exhibit, I wandered outside and walked down a dirt path to a Quonset hut. I walked in and saw a horrific sight—tables piled with skulls, remains of those who had perished in nearby woods and swamps. Machetes were the primary instrument of death in the genocide and I could see machete marks in the skulls. The horror was overwhelming and I began to sob.

Alice had followed me into the building. She gently put her arms around me and led me outside. "You don't have to look at this," she said. "This isn't *your* problem; this is *our* problem."

I thought about those words for a long time.

I returned home. A few months later, I was teaching an introductory sociology class and we were studying the Rwandan genocide. A student approached me after class one day, upset by what we had been discussing, and she asked, "What are we going to *do* about this?"

I thought I'd been *doing* a lot—teaching about genocide, encouraging my students to think about global issues, and working with refugees from Africa. That question, though, challenged me to make a direct difference.

I realized that genocides happen for many complicated reasons and also for a simple one—we let them happen, we are bystanders. My life had brought me to this *enough moment*, a moment when I had to take a stand and *do* something. Genocides aren't natural disasters like tsunamis or hurricanes or earthquakes. Genocides are caused by human beings, and therefore can be prevented.

We have examples of successful *social movements*[1] that have solved almost insurmountable problems in the past. Today's anti-genocide strategies are the legacy of England's movement to abolish slavery in the 19th century, the American civil rights movement in the 20th century, and an international movement for peace and justice that began after the Holocaust.

Like nearly all social movements, today's powerful anti-genocide citizen movement began with a *precipitating event*. In 2004, the U.S. government labeled the conflict in Darfur to be genocide, the first time in history that the United States has given a crisis that name while it was actually occurring. The year 2004 was also the 10th anniversary of the Rwandan genocide, marked by the release of films chronicling global inaction in the face of unspeakable tragedy and with world leaders' visits to Rwanda and efforts to apologize for actions not taken.

The juxtaposition of this apology for Rwanda and the simultaneous declaration of a new genocide in Darfur produced moral outrage. A new social movement to prevent genocide, in many ways more energized and more focused than earlier efforts, began to develop. This illustrates the first stage in the development of a social movement, *emergence*; people are upset with a situation and organize to voice their grievances.

Individuals began to raise awareness about Darfur and to form groups for action. Anti-genocide activists created the Save Darfur Coalition,

---

[1]Information on social movement processes is taken from Donatella De la Porta and Mario Diani, *Social Movements: An Introduction* (Blackwell, 2000).

Genocide Intervention Network, Stop Genocide Now, STAND, and the Enough Project between 2004 and 2007. Existing human rights organizations, such as Amnesty International and Human Rights Watch sounded the alarm as well. The "Save Darfur" movement created a base in colleges and faith organizations throughout the country, a mobilizing strategy used successfully during the civil rights and anti-apartheid movements. This was the second stage, *coalescence*; the organizations, through the emergence of strong leaders, began to work together to plan events and generate political action.

Key world leaders were also taking action. In 2001, a Canadian-led commission concluded that the world has a "responsibility to protect" innocent people when their own governments are either unable or unwilling to do so.[2] This mandate, known as R2P, redefines a nation's obligation to its people to protect them from being targeted based solely on who they are—by their race, religion, national origin, or ethnic identity. R2P holds that the world must intervene in the face of humanitarian crises.

Why did Canadians lead the effort to enact R2P? Lloyd Axworthy, a Canadian, served on the Commission that created the R2P mandate; he had twice been the president of the United Nations (UN) Security Council, was a strong advocate for multilateralism in foreign affairs, and was nominated for the Nobel Peace Prize. Another Canadian, General Romeo Dallaire, had led the ill-fated United Nations peacekeeping mission in Rwanda during the genocide. He spoke out widely about the tragedy he had witnessed firsthand, and become an impassioned advocate for peace.

After the Commission's report, the UN passed a similar resolution: there is a responsibility to protect.[3] Soon after, a U.S. committee developed a blueprint for genocide prevention.[4] The failure to protect genocides in Rwanda and in Bosnia in the 1990s had become global calls to action from the top down, including the United Nations, and from the bottom up, including hundreds of student groups across the country. The anti-genocide movement became *bureaucratized*, the third step in a social movement. The organizations hired experts, moved from mass rallies and charismatic leaders to organized political action for social change, and developed strong collaborations and coalitions among sectors of civil society.

---

[2]*The Responsibility to Protect, Report of the International Commission on Intervention and State Sovereignty*, International Development Research Centre (Canada, December 2001).
[3]Unanimously endorsed at the United Nations General Assembly World Summit, September 2005 and reaffirmed by the United Nations Security Council, April 28, 2006, in Resolution 1674.
[4]Madeleine K. Albright and William S. Cohen, *Preventing Genocide: A Blueprint for U.S. Policymakers*, with the United States Holocaust Memorial Museum, The American Academy of Diplomacy, and the Endowment of the United States Institute of Peace (2008).

Decline is the final step in a movement, either because of a failure to develop appropriately, or, hopefully, because the movement has been successful and the desired change has happened, such as women getting the right to vote. The anti-genocide movement has not yet reached this point; genocides still happen. The increasing strength of the movement suggests that someday we may, indeed, have a world without genocide. We can make that our legacy just as previous generations, through their social movements, have legally abolished slavery and guaranteed women's rights in many parts of the world.

It was within this context—a growing awareness of the first genocide of the 21st century, shifting global norms mandating protection for innocent civilians, and a proliferation of organizations that, by 2005, were already bureaucratized, visible, and credible—that my student asked me that question, "What are we going to *do* about this?"

My own story had pointed me in this direction for a long time. I'm Jewish. I grew up in the aftermath of World War II with nightmares about being targeted like Anne Frank. My parents had friends who survived the Holocaust—and relatives who did not. I've been to Auschwitz, Cambodia's killing fields, Hiroshima, East Timor, Rwanda, and Bosnia. I was ready to act. My student's question was my "precipitating event" and my "enough moment." I began the organization World Without Genocide in 2006, shortly after going to Rwanda and after my student challenged me with her question.

I am a *political sociologist*[5], using sociology for political ends. World Without Genocide's mission is to protect, prevent, prosecute, and remember. We advocate for measures to *protect* innocent people across the globe from being targeted because of their race, religion, ethnic or national origin, or sexual identity. We support efforts to *prevent* discrimination. We advocate for the *prosecution* of perpetrators and an end to impunity. And we believe we must *remember* those whose lives and cultures are affected by violence.

As an organization, we're now in the third phase, bureaucratization. We have a staff, we are headquartered at a law school, and we work within the political arena at the state and national levels. We focus on education as the first step toward change. We teach about genocides, mass atrocities, human and weapons trafficking, child soldiers, and gender-based violence. We hold conferences; we made a film about children during genocide[6]; we launched the country's first institute on genocide for high-school students;

---

[5]Michael Burawoy, past president of the American Sociological Association, has done a great deal to expand the profession's engagement with this domain of the discipline.

[6]*Children of Genocide: Five Who Survived,* featuring interviews with survivors of the Holocaust and the genocides in Cambodia, Bosnia, Rwanda, and Sudan; nominated in 2010 for a regional Emmy award; available by contacting info@worldwithoutgenocide.org

and we have student chapters of our organization. We run a large internship program to engage the next generation in human rights.

As Mark Hanis, founder of the Genocide Intervention Network[7], says, "Knowledge is not power. Knowledge plus *action* equals power." Therefore, we also work to change laws. In 2007, we helped pass a bill to divest Minnesota's pension funds from companies complicit with the Darfur genocide. In 2011, we spearheaded the successful effort to have April designated as Genocide Awareness and Prevention Month in Minnesota. We support the national campaign to end the use of *conflict minerals* in small electronics, a practice that fuels violence in the Democratic Republic of the Congo. We advocate for the United States to join other nations in ratifying the Convention on the Rights of the Child, the Convention to End Discrimination Against Women, and the International Criminal Court.[8]

We educate. We get laws passed. We tell survivors' stories. And we create leaders. I invite you, future sociologists, to join us in creating this legacy, a future world without genocide!

CHANGE THE WORLD—START AT SCHOOL

Mikaila Mariel Lemonik Arthur

Rhode Island College, Providence

Mikaila Mariel Lemonik Arthur is an assistant professor of sociology at Rhode Island College, where she researches social movements and curricular change and teaches courses on research methods, law and society, and race. In graduate school, she was an activist with the New York University graduate employees union, and as an undergraduate at Mount Holyoke College she used sociological research to help convince the administration to build a kosher/halal dining hall. Her book, *Student Activism and Curricular Change in Higher Education* (Ashgate, 2011), draws on the research discussed in this chapter.

---

[7]The Genocide Intervention Network merged with the Save Darfur Coalition in 2011 to form United to End Genocide.

[8]The United States is one of only two member states in the United Nations that has not ratified the Convention on the Rights of the Child and one of only eight that has not ratified the Convention to End Discrimination Against Women. At this writing (fall 2011), 118 nations have ratified the International Criminal Court; the United States is not one of them.

When I was an undergraduate at Mount Holyoke College, at the turn of the new millennium, I took a course in Asian American studies. In the course, I learned that students had been pushing for the college to adopt an Asian American studies major since 1997, when students had occupied a building to demand increased attention to diversity on campus. While the college had offered degrees in African American studies for a long time, the course I was enrolled in was one of the very first that had been offered in Asian American studies. This really got my attention—why was Asian American studies not a bigger part of the educational landscape? Had those students who occupied the building several years earlier made a difference in the curriculum on our campus?

These questions have continued to follow me over a decade. In the semester after my Asian American studies course, I took a sociology course focusing on social movements. For my final term paper, I studied the Asian American civil rights movement of the late 1960s. Though this movement had many other goals, one of the places where it really crystallized was around the effort to establish Asian American studies on college campuses in California. One key episode in this early history of Asian American studies was a student strike at San Francisco State College in 1968–1969 that closed down the campus for several weeks. My paper was really my first attempt to use sociological theory to empirically examine the social situation I observed.

As I started to develop a dissertation project in graduate school, I realized that the story of Asian American studies was not unique. A variety of other new intellectual fields were emerging in the late 1960s and early 1970s alongside Asian American studies. Students, faculty, and staff across the United States had put their educational and career success on the line to advocate for the inclusion of new fields of knowledge in the college curriculum. They held protests, occupied buildings, organized their own courses outside the regular curriculum, and even risked arrest because of their belief that a place should be made inside the curriculum for the disciplines they wanted to study. I decided that my dissertation research would examine how it was that three such disciplines found their way into American colleges and universities—Asian American studies, of course, but also women's studies and queer/LGBT studies.

The process of forming new curricula in colleges and universities has two main parts. First, new areas of study must become part of the knowledge environment. In other words, areas of study that once did not exist must be created. Before the 1960s, people simply did not do research on or teach about issues related to women, Asian Americans, or queer/LGBT people. The emergence of these new disciplines has led to broad changes in American academic and social life. Just to take one example, before women's studies became a part of the curricular landscape, authors and publishers found nothing wrong with publishing sociology textbooks on marriage and the family without providing any coverage of women's issues. Today, due to the social movements that brought this issue forward, a textbook on marriage and the family that ignored women's roles and experiences would be unlikely to find a mainstream publisher. The second part of forming new curricula is that advocates for new areas of study must convince individual colleges and universities to adopt these programs of study.

To answer my research questions, I selected six diverse colleges and universities across the United States. They varied in terms of size, location, public or private status, and selectivity. I visited all six institutions and spent several days in each campus's archive, locating documents that told the stories of efforts to establish programs in Asian American studies, women's studies and queer/LGBT studies. While in the archives, I developed lists of students, faculty, and staff who had been active in the efforts to establish these programs and then sought them out. I made phone calls to as far away as Australia (to find a faculty member on sabbatical) and Norway (where a graduate of one of the colleges in my study had moved to start a family). Some of the people I contacted had not thought of their participation in these campaigns in years and were thrilled that someone was finally interested in their stories. Others were unwilling—or at least extremely reluctant—to revisit the past. Ultimately, I interviewed over 100 individuals who had taken part in or observed one or more curricular change campaigns.

Some of the movements I studied were successful, and their efforts resulted in the creation of programs or departments that remain vital parts of their campuses. Others were not so successful, leaving campuses without such programs even several decades later. In still other cases, no activists ever emerged to try to start a program in the first place! My research question thus ultimately focused on what explained these differences. What factors made it possible for curricular change campaigns in some instances to result in successful, institutionalized programs while others were barely a blip on the radar screen?

Sociologists who had studied social movements in the past had proposed a variety of explanations for what factors make some movements more able to have an impact than others. For instance, Edwin Amenta, a researcher who studies social movements that target the U. S. government, has argued that movements that are able to match their strategies to the political context in which they find themselves will be most able to have an impact. Where government officials seem unfavorable and inaccessible to a movement's demands, movements will do best if they choose assertive strategies like protests, while in situations where government officials are open and favorable to their demands, movements will do better to choose more mainstream strategies (Amenta, 2006; Amenta & Caren, 2004; Amenta, Halfmann, & Young, 1999; Amenta & Young, 1999). But these previous researchers had not developed theories designed to explain the impacts of movements targeting organizations like colleges and universities. Rather, their models were designed just for the political sphere, a somewhat related but not directly analogous realm. I therefore had to develop my own model for understanding what made the movements I studied capable of having an impact.

Social movements targeting colleges and universities, I found, were more able to have an impact when they adjusted their strategies to fit the degree of openness and flexibility of the administration as well as the favorability of the college or university's mission to the types of changes the movement proposed. The strategic choices that matter most are those that have to do with how social movements talk about themselves and their goals ("framing") and the extent to which they position themselves as insiders or outsiders in relation to their college or university. Unlike in the case of movements targeting the political sphere, movements targeting colleges and universities find that their strategies must be *less* assertive the less favorable the context is.

This theoretical model is important as a new way for sociologists to understand social movements that target organizations. To disseminate my findings to other academics and researchers, I have presented the results of my research at a number of sociological and interdisciplinary conferences and have published in peer-reviewed journals. But my research also provides a new perspective for campus activists on how to make social change happen inside colleges and universities. Since completing my project, I have worked to ensure that people who are interested in creating organizational change within their own colleges and universities will have access to my findings as part of a toolkit of ideas and techniques they can use in their own campaigns.

In 2008, a group of Mount Holyoke College students who were determined to renew the push for Asian American studies contacted me. At the time of this writing, Mount Holyoke College, where I first learned about Asian American studies, still does not offer a major or minor in the field. In the 2009–2010 academic year, the college offered only two courses—both in literature—focusing on the Asian American experience. However, students can earn a certificate in Asian American studies by taking classes at other area colleges, and the group of students who contacted me wanted to learn more about the history of this certificate program and efforts to create Asian American studies at Mount Holyoke so that they could build on this history in their renewed activism. Because of the turnover among students, social movements on campus often have limited institutional memories. In other words, they are unable to keep track of what has been done before, why those decisions were made, and what their consequences were. My research helped the Asian American student activists at Mount Holyoke build an institutional memory that would serve as a foundation for their future plans. In the months following our conversation, student activists held a celebration recognizing a decade of Asian American student activism at Mount Holyoke.

I have disseminated my research findings to broader audiences as well—there does not have to be a divide between research that makes a scholarly impact and research that makes the world a better place. I have spoken about strategy building for social movements within organizations at conferences for academics as well as at events for teachers and students interested in making curricular change in their own schools. I have also published articles on curricular change strategies in academic journals (Arthur, 2008b, 2009) as well as in nonacademic venues, such as a publication for people looking to make their colleges and universities a better place for women (Arthur, 2008a). Making use of these findings can help reduce the burden created by the lack of institutional memory on individual campuses so that social movements interested in making change can begin to do so with a greater chance of making an impact.

I am, of course, also writing about my research for you. So what will you do with these findings? How can you make your own institution a better place?

References

Amenta, E. (2006). *When movements matter: The impact of the Townsend Plan and U.S. social spending challengers.* Cambridge, MA: Russell Sage Foundation.

Amenta, E., & Caren, N. (2004). The legislative, organizational, and beneficiary consequences of state-oriented challengers. In D. A. Snow, S. A. Soule, & H. Kriesi (Eds.), *The Blackwell companion to social movements* (pp. 461–488). London, UK: Blackwell.

Amenta, E., Halfmann, D., & Young, M. P. (1999). The strategies and contexts of social protest: Political mediation and the impact of the Townsend movement in California. *Mobilization, 4,* 1–24.

Amenta, E., & Young, M. P. (1999). Democratic states and social movements: Theoretical arguments and hypotheses. *Social Problems, 57,* 153–168.

Arthur, M. M. L. (2008a). Activist strategies to change your campus curriculum. *Women in Higher Education, 17,* 34–35.

Arthur, M. M. L. (2008b). Social movements in organizations. *Sociology Compass, 2,* 1014–1030.

Arthur, M. M. L. (2009). Thinking outside the master's house: New knowledge movements and the emergence of academic disciplines. *Social Movement Studies, 8,* 73–87.

A CAMPUS GUN-FREE ZONE MOVEMENT

Rob Benford

University of South Florida, Tampa

Rob Benford received his PhD in sociology from the University of Texas at Austin. He is a professor and chair of the sociology department at the University of South Florida in Tampa. He has written extensively on the peace, environmental justice, and college sports reform movements. He previously served as editor of the *Journal of Contemporary Ethnography* and as the series editor of Twayne Publishers' *Social Movements Past and Present* series. He also has served as president of the Midwest Sociological Society, and as chair of the Peace, War, and Social Conflict and the Collective Behavior and Social Movements sections of the American Sociological Association.

One trait shared by most sociologists is that our sociological imaginations tend to operate 24/7. Thus, we tend to interpret the world around us sociologically—understanding personal troubles or community concerns in terms of larger social patterns and forces. Occasionally, we decide to take action to try to affect local problems or perceived threats to our communities by altering the social forces we believe are at the heart of our troubles. That's precisely what fellow sociologist Jack Siegman and I did in the mid-1990s when we mobilized a midwestern university to create a campus gun-free zone.

From the fall of 1992 to the fall of 1994, a series of events unfolded on or near the University of Nebraska-Lincoln (UNL) campus that amplified our fears concerning the proliferation of guns and thus the increased chances of gun violence. The first and most dramatic incident occurred on October 12, 1992, when a graduate student walked into his actuarial science class, pointed a semiautomatic assault rifle at students, and pulled the trigger. Fortunately, the gun jammed; as he slammed the rifle's butt against a desk trying to unjam it, the students managed to pin him against the wall with a desk and escape. "We all went to the floor, pulled the desks over our heads," recalled one student. Another said, "I remember the guy pointing the gun at all of us and then it was just a mad rush—desks flying. We were really extremely lucky." The attempted killer, who had two 30-round clips of ammunition, was later caught, arrested, and convicted of attempted second-degree murder but found not responsible by reason of insanity. He remains in state custody.[9]

I recall thinking how lucky we were as a campus community that the would-be mass murderer's gun had jammed. However, several more incidents involving guns on or near the UNL campus made me wonder how long our luck would hold. A year after the near massacre, a student flashed a weapon at a campus police officer while driving his car on a road adjacent to campus. When the officer attempted to apprehend him, the student wounded the officer. After the student's arrest, campus officials revealed that campus police had previously confiscated the same weapon from his dorm room on the grounds that he had violated university housing's prohibition against having guns in the dorms. However, citing the Second Amendment, a judge ruled that UNL officials were required to return the gun to him.

A few months later, a third gun-related incident involving a student heightened our growing concerns. A University of Nebraska football star was arrested near campus for firing two shots at another moving vehicle from the car he was in. The shooting was a continuation of an earlier dispute involving college and professional football players that had begun at an off-campus party. The shooter pleaded no contest to a felony charge of unlawfully discharging a gun and to misdemeanor assault, served a few months in the Lancaster County Jail, and went on to star in the National Football League.[10]

Whereas the three gun incidents sparked my sociological imagination, it was the results of systematic data gathering that prompted me to take action. In the spring of 1994, Kelly Asmussen, a student pursuing his doctorate, conducted a survey in my two large introductory sociology classes (the survey was approved by the institutional review board and the participation of my students was voluntary). A few months later, Kelly provided me with the general results from his survey. I was stunned at what the data revealed and their implications. Two survey items in particular caught my attention:

1. How often have you kept any of these items of protection: gun, knife, gun other than a handgun, somewhere on this campus (such as in your car or your place of residence)?

    (A) Never   (D) Often
    (B) Once   (E) Very Often
    (C) A few times

---

[9]Quotes are from a 2007 KETV (Omaha) broadcast downloaded on February 19, 2010, retrieved from http://www.ketv.com/news/12199124/detail.html. For a description of the incident, see Kelly Asmussen and John W. Cresswell, "Campus Response to a Terrorist Gun Incident" (*Journal of Higher Education*, 66(5), 575–591, 1995).

[10]Green Bay Packers' Tyrone Williams Sentenced to Six-Month Jail Sentence. (*Jet*, Dec 9, 1996). Available at http://findarticles.com/p/articles/mi_m1355/is_n4_v91/ai_18949541/

2. How often have you carried a knife or gun with you while you attended classes on this campus?

(A) Never   (D) Often
(B) Once   (E) Very Often
(C) A few times

Although the raw data Kelly provided me with did not include a breakdown of the frequency of the activities, I was shocked to learn that 14.2% of the male students and 4.6% of the female students admitted that they had carried a gun or knife to campus. Although the percentage indicating they had actually carried such weapons to classes was smaller—6.9% for males and 1.0% for females—I was even more disturbed by those reports. I did the math. Given that approximately 300 students (of 350 enrolled) attended each of my two sections of introductory sociology each Monday and Wednesday morning, I could expect that there may be a dozen weapons present in each class. It occurred to me that playing poker in a Wild West saloon might have been a safer vocation!

As I thought about the shooting incidents and Kelly's disturbing findings over the summer break, I decided to float the idea of a "gun-free zone" among my colleagues. I borrowed the idea from my studies of the nuclear disarmament movement in the 1980s. A number of small island countries, as well as a few municipalities around the world, had declared their jurisdictions to be "nuclear-free zones."[11] Why not apply the same logic to our campus and not allow guns within UNL's borders?

When classes resumed in the fall of 1994, I first approached my friend and colleague Jack Siegman. I recruited Jack to help me get the gun-free zone idea going for three reasons: (1) I trusted him, (2) he had extensive social ties throughout the UNL community, and (3) he had a reputation for challenging the status quo. Drawing on our knowledge of social movements and complex organizations, Jack and I decided that in order for the gun-free zone movement to be successful, we would need to mobilize campus leaders, including not only UNL administrators, but also representatives of various constituency groups on campus. We subsequently held a series of meetings with representatives and directors of student affairs, housing, police, judicial affairs, Greek affairs, student government, and the faculty senate, as well as office professionals and professors representing several departments.

---

[11]Robert D. Benford, "The Nuclear Disarmament Movement," in L. R. Kurtz (Ed.), *The Nuclear Cage: A Sociology of the Nuclear Arms Race* (Englewood Cliffs, NJ: Prentice Hall, 1988), 237–265; David C. Pitt and Gordon Thompson (Eds.), *Nuclear-Free Zones* (London, UK: Croom Helm, 1987).

One of the first things we realized was that not everyone agreed that guns on our campus constituted a social "problem." Hence, we spent time in our initial meetings convincing others that we were sitting on a powder keg and that they should join our gun-free zone movement. Fortunately, we enjoyed immediate support from most campus leaders, perhaps most significantly from UNL's police chief, Ken Cauble. The chief articulated the magnitude of the problem in an initial meeting:

> We have seen an increasing number of guns, especially handguns, on campus lately . . . [in the possession of] students and non-students. It's almost a nightly deal where we receive reports of shots fired. . . . We are also fielding an increasing number of requests from parents wanting mostly daughters to be allowed to carry guns to and from home.

Chief Cauble went on to explain that among the items most frequently reported stolen from students' vehicles were guns, including hunting rifles students stored in their cars and trucks. Thus we learned from the chief that we should not focus just on students (but should also include visitors to campus), that there were gendered dimensions to the problem (with some parents wanting their daughters armed for protection when traveling to and from home), and that there could be rural–urban (hunters versus non-hunters) subcultural differences at work—all of which we then sought to address in our policy recommendations.

First, we surveyed other Big Eight conference schools regarding their firearms policies and used this information to help write a new student policy for UNL. We recommended that the student code of conduct be revised so as to not permit guns on campus and, in the interests of fairness and equality, we sought to extend the policy to *all* members of the UNL community, including faculty and staff. After considerable debate, our recommendations were adopted by UNL's student government and the Board of Regents, and were put into effect the following fall. A year later, the gun ban was extended to faculty and staff.

Second, we dealt with the gendered aspects of safety fears and personal security. Several women from across campus ranks (students, staff, faculty) told us they carried guns in their purses for protection. We forwarded recommendations to the administration that they address these genuine concerns by enhancing services, such as campus escorts, transportation, additional lighting in parking lots and along walkways, emergency kiosks, and programs that deal with acquaintance rape. What we sought to do was to decrease the need for women to arm themselves with lethal weapons.

Third, it became apparent not only from Chief Cauble's remarks but also from a variety of quarters, that we should address the needs of hunters. Some pro-gun folks argued that we were trying to take their guns away and that we were ignoring hunters' rights to pursue their preferred recreational activities. "Some people think guns are here only for bad reasons. That is not the case," argued one student. "There are a lot of hunters here," he added.[12] Chief Cauble volunteered a solution. He offered space in the police station where hunters' guns could be kept in lockers. This would get the weapons out of students' vehicles and decrease the likelihood of accidental shootings, as well as shootings carried out by persons under the influence of alcohol or drugs, or in a fit of anger. The gun locker option also supported our claim that we were not seeking to revoke the Second Amendment. The gun locker proved to be quite popular. One year after we started our gun-free zone movement, 54 lockers were full of guns and ammunition and the police had expanded the program. "Originally, we had only one set of lockers, but there was such a demand that we purchased more," Chief Cauble observed a few months after the policy went into effect. "We really didn't think the system would work out this well," he added.[13]

Although Jack Siegman and I had been involved in various attempts to affect social change throughout our careers as sociologists, including the civil rights, anti-apartheid, peace, environmental, and college sports reform movements, we found our experiences with initiating and mobilizing a local gun-free zone movement to be among the most gratifying. This was due in part to the fact that we were able to focus our sociological imaginations on a specific problem and achieve nearly immediate, observable results.

SOCIAL MOVEMENTS AND ACTIVIST SOCIOLOGY

Charles Derber

Boston College

Charles Derber is professor of sociology at Boston College and has written fifteen books, reviewed in the *New York Times*, the *Washington Post*, the *Boston Globe*, and other leading media. His books, which include *Corporation Nation, The Wilding of America, The Pursuit of Attention, People Before Profit, Greed to Green, Marx's Ghost*, and *The Surplus American*, and *An Invitation to Political Economy* have been translated into five languages. He has also written for the *International Herald Tribune*, the *Boston Globe, Newsday, Tikkun*, and many other periodicals. Derber is a life-long activist who is engaged in peace, environmental, labor, and other social justice movements, including Occupy.

---

[12]Betty VanDeventer, "Firearm-Free UNL Campus Likely" (*Lincoln Journal-Star*, March 11, 1995), 1.

[13]Jason Levkulich, "UNL Police Gun Lockers a Popular Draw for Students" (*The Scarlet*, University of Nebraska-Lincoln, March 15, 1996), 1.

Activism and sociology are intertwined like DNA in my life story. The summer before I decided to go to graduate school in sociology at the University of Chicago, I worked registering African-Americans to vote in Mississippi, experiencing the clash between the armed world of Southern segregationists and the courage of Black civil rights activists. I was taken hostage by some White racists, shot at on a country road when driving with Black activists, and slapped in jail by some White sheriffs with no love for Yankee boys. Going south for civil rights work was my first form of "sociological activism," and I learned more sociology than I had in my entire college career: about race relations, state violence, and social change. I henceforth always saw personal participation in social movements as a moral and intellectual necessity.

There are many types of sociological activism but mine is all about building social movements. This is not based on nostalgia for the 1960s (although I have some) but on a sociological conviction that social movements are unique—the only social organizations capable of creating major democratic systemic change. Historically, the abolitionist, suffragette, and labor movements have been leading champions of social revolutions. Building social movements is the only true survival strategy for all citizens in a world facing climate change, rampant militarism, and predatory corporate capitalism, but it is a special responsibility for sociologists whose goal is to analyze social systems and help change them.

This is not to say that sociologists should be propagandists for particular movements. Rigid or "politically correct" thinking makes for terrible pedagogy and anemic activism. Sociological activists should never be in the business of trying to police a politically correct worldview or coerce activism from their students, a violation of both their academic authority and the spirit of critical inquiry that allows both sociology and social movements to succeed.

Sociological activists are typically passionate about both their ideas and their commitments to social change. That makes for good sociology and good social movements. Sociologists should, in fact, acknowledge their values and passionate commitments to movements and justice, a form of honesty that leads to trust and more authentic dialogue on and off campus. There is rarely much learning or action without passion, as long as it does not get converted into propaganda, in which case passion becomes blind and can inhibit necessary exposure to a wide range of clashing theories—including those critical of one's own political worldview.

There are many models of movement-oriented sociological activism, one being the use of one's knowledge and work-based resources to help organize new organizations and visions within social movements. In 1971, after I joined the Brandeis sociology department as an assistant professor, I helped organize a National Student Strike Center against the war, coordinating work around the country to help mobilize all-out resistance to President Nixon's expansion of the war in Indo-China. The Student Strike Center created branches on campuses across the country, coordinating national anti-war protests and helping reframe the anti-war movement as a struggle against global capitalist control. The Strike Center brought together anti-war activists from different campuses and connected them with other movement groups, such as the Black Panthers, a militant anticapitalist and anti-Vietnam African American organization.

As a professor, I led a faction of "soft activists" seeking dialogue and participatory consensus, while my faculty office mate led a "hard activist" wing, stressing discipline and doctrine. We were both sociological activists but we fought about vision, strategy and tactics, my faction always arguing for inclusiveness, multiplicity of visions, and nonviolence. Sociologist Alvin Gouldner (1980) described "Two Marxisms": the humanist, antiauthoritarian and free-spirited ideals of the young Marx versus the scientific, authoritative and deterministic Marx of later years. These are metaphors for divisions in many justice movements—both Marxist and non-Marxist. Despite the "two Marx" divisions in the National Strike Center, it became, for a short period, a sociological nerve center of the national antiwar movement, and affirmed my view that sociologists could help mobilize and reframe movements from the inside.

Another activist sociology chapter of my life began with the 1999 Battle of Seattle, where labor, peace, and environmental activists converged in the streets of Seattle to oppose neoliberal corporate globalization. I had been writing about globalization in several books getting public attention—part of a sociological activist career is writing for the public to which social movements must appeal. With the students who accompanied me to Seattle, I founded a new organization, the Global Justice Project (GJP), a campuswide social justice movement at my new academic home, Boston College.

The best thing the GJP did was partner with the National Labor Center—a leading global antisweatshop organization—to bring young female sweatshop workers from Bangladesh and El Salvador to campuses in the Boston area. These girls, often illiterate, told their heart-wrenching stories of working 20-hour days and sleeping under their sewing machines to the American students of their own age, who openly wept, with emotion so raw that it is still vivid in my memory. GJP helped them turn grief into activism. Scores of U.S. students worked with GJP to put pressure on the global corporations and force campuses to insist on worker rights. They sent petitions to CEOs to end sweatshop practices and forced disclosure of their own university's investments, insisting on sweat-free caps and sneakers in their campus stores. They demanded their campus join monitoring efforts of factories in poor countries, such as those carried out by the Worker Rights Consortium, a global network of labor and human rights activists. Many built long-term relations with the sweatshop workers they couldn't forget.

GJP still exists, and has become one of the student groups helping form the Occupy Movement today, demonstrating that sociological activists can put in place enduring organizations for change. After the financial meltdown of 2008, a new national wave of resistance emerged. In my books of this period, such as *Greed to Green* (2010) and *Marx's Ghost* (2011), I argued that the fragmented identity movements of gender, race, and labor must unite to confront the intertwined triple crises of capitalism, climate change and militarism.

At just this moment, Boston-area life-time activists in these very movements were coming to the same conclusion. We created together a new organization called the Majority Agenda Project (MAP) to help build new organic ties between movements running on separate tracks. In 2009 and 2010, MAP brought together labor leaders with climate and peace activists. MAP helped nurture creative collaborations, such as One Nation, a national movement for capitalist transformation led by the national SEIU (Service Employees International Union) and the NAACP. I'll never forget the huge protest held on October, 2010, on the Washington Mall, led by African American unionists. What a sight to see poor Blacks and Whites in their union shirts walking and singing together for a "new nation," evoking Martin Luther King Jr.'s dream of a multiracial movement for peace and justice. Unions such as SEIU began leading social movement struggles not just for high wages for their members but national protests to protect Social Security, Medicare, Medicaid, and union rights.

In 2011, the Occupy Wall Street movement was born. Its immediate predecessor was the Wisconsin Wave, where thousands of workers, students (including sociology graduate students at the University of Wisconsin, housed right next to the Capitol), farmers, and environmentalists occupied the Wisconsin State House to protest Republican Governor Scott Walker's move to eliminate public sector unions. Shortly thereafter, the Occupy Movement exploded on the scene, as activists set up tents in the shadows of the biggest banks in the world. I felt like I had been waiting for Occupy my whole life. It proliferated like spring flowers sprouting through cracks in the sidewalk in every American city, creating a diverse, colorful and spontaneous community of activists from every progressive movement, now uniting to confront the Wall Street tycoons running the country.

Occupy became my sociological and activist preoccupation. I wrote *Marx's Ghost* (2011) and *The Surplus American* (2102), two books that offered analysis and theatrical productions for explaining and building the movement. I worked with MAP to offer help to young Occupy activists, sharing history of previous Occupy movements and our thoughts about where to go from here. I devoted entire classes to exploring the economic crisis and how Occupy might be the movement that could unite all the diverse identity groups, and bring a new generation of Americans into critical thinking and visionary political activism.

Most sociologists will become better scholars and teachers if they engage in sociological activism. My best writing has been animated and informed by my participation in real world struggles to change socially dominant and violent institutions. Movement participation is one of the best sociological methodologies for collecting data and building theory.

It is thus hardly surprising that some of the best sociologists, often without academic credentials, are activists who work in the community and outside the academy. Their activism is enriched by their sociological analysis, and these "organic intellectuals" often teach their academic counterparts. The dialogue between these two communities of sociological activists—and their mutual involvement in grass roots movements—not only is the lifeblood of good sociology but one of the great forces of social transformation.

Existential social crises require activist sociologists. The clock is ticking and we need to be in the trenches. It can be risky and time-consuming but also a source of wonderful connections and deep meaning. If we don't translate our sociological knowledge into action, sociology will fail and society will too. Several years ago, students who watched me get arrested with several other sociology faculty while committing civil disobedience in front of a big bank told me it was the most educational moment of their college experience. Our own actions will always speak more eloquently than our words.

References

Derber, C. (2010). *Greed to green: Solving climate change and remaking the economy*. Boulder, CO: Paradigm Publishers.

Derber, C. (2011). *Marx's ghost: Midnight conversations on changing the world*. Boulder, CO: Paradigm Publishers.

Derber, C., & Magrass, Y. R.. (2012). *The surplus American: How the 1% is making us redundant*. Boulder, CO: Paradigm Publishers.

Gouldner, A. (1982). *The two Marxisms: Contradictions and anomalies in the development of theory*. New York, NY: Seabury Press.

1. How will you respond to Ellen Kennedy's invitation to join her in "creating a future world without genocide"? Why? What sociological tools will you bring with you to address this challenge?

2. Kennedy describes how nearly all social movements begin with a precipitating event. What event or events sparked the anti-genocide movement? Has there been an event that has prompted you to create or join a social movement? If yes, why? If no, why not?

3. Mikaila Mariel Lemonik Arthur points out that changes in college curricula have impacted how groups such as women and other minorities are perceived and treated. Why do you think it is important to have courses that focus on issues of (a) gender, (b) race and ethnicity, and (c) sexuality?

4. At the end of her Sociologist in Action piece, Arthur asks, "How can *you* make your own institution a better place?" Regarding how underrepresented groups are perceived and treated on your campus, how might you answer her? Be specific and thoughtful.

5. Is your campus gun free? Why do you think it is (or is not)? What interest groups or events on your campus have influenced its becoming (or not becoming) gun free? Would you be interested in making yours a gun-free campus (if it is not already)? Why or why not? If yes, how might you go about creating this change?

6. How did Rob Benford and his colleagues draw on their knowledge of social movements as they began their efforts to start a gun-free social movement on their campus? What do you think is the key to a successful social movement? Why?

7. Charles Derber maintains that "if we don't translate our sociological knowledge into action, sociology will fail and society will too." What do you think he means by this? Do you agree? Why or why not?

8. Derber points out that "there is rarely much learning or action without passion." Provide examples from your own life that prove this point.

9. Describe one thing you learned from each of the Sociologist in Action pieces about forming a successful social movement. Which Sociologist in Action piece do you think would be most helpful for those interested in starting a social movement? Why? Does it depend upon the particular movement? Why or why not?

10. Each of these pieces somehow relates higher education to social movement work. Why do you think institutions of higher education are so often connected to social movements? Consider a social change you would like to see in the world. How might you begin a social movement on your campus that could ultimately influence the larger society?

RESOURCES

The following Web sites will help you to further explore the topics discussed in this chapter:

198 Methods of Nonviolent Action	http://www.aforcemorepowerful.org/resources/nonviolent/methods.php
ASA Section on Collective Behavior and Social Movements	http://www2.asanet.org/sectioncbsm/
Campus Activism	http://campusactivism.org/
People Power	http://www.peoplepowergame.com/
Sociosite Social Movements	http://sociosite.net/topics/activism.php

To find more resources on the topics covered in this chapter, please go to the Sociologists in Action Web site at **www.sagepub.com/korgensia2e.**

Chapter 8

Stratification and Social Class

There is no avoiding it. *Stratification*, the ranking of groups according to their access to and possession of what is valued in society, happens everywhere. Whether a society is capitalist, communist, or a dictatorship, it has a system for making sure that some people have more of what is valued while others have less. In the United States, one of the most important ways we rank people in our system of stratification is according to their social class level (upper, middle, lower).

One of the goals of sociology is to provide a voice for the marginalized, those who have the least power in their society. In this chapter, three Sociologists in Action describe the ways they have done just that. Using sociological tools, they each help to provide those on the lowest levels of the stratification system with the means to gain more power.

Joe Bandy and Craig McEwen begin the chapter with "Housing and Homelessness in Maine: A Case of Public Sociology in Practice." In this piece, they describe how they formed research partnerships with local agencies working to end homelessness and to provide affordable housing in the larger community surrounding Bowdoin College. While learning research methods, their students carried out interviews that gave voice to those waiting to receive housing assistance and those residing in homeless shelters. They also conducted opinion polls of attitudes toward homeless people in Maine and other research that helped a local housing organization "develop affordable housing in ways that would mitigate public misconceptions and fears associated with homelessness." In the process, the students learned about the housing crisis in Maine and the power of sociology to address social issues.

In the second piece in this chapter, "Relocating the Homeless—or Not!" James Wright relates how he and his graduate students used sociological tools to understand the potential impact of the proposed relocation of the Coalition for the Homeless of Central Florida, the region's largest provider of homeless services, from a traditionally African American neighborhood in Orlando, Florida, in an effort to "gentrify" the area. Wright and his students conducted this research to "bring the voices of homeless people themselves into this debate" and make sure that the various possible repercussions of the move were considered during the decision-making process. Without the work of Wright and his students, it is highly likely that the proposal to move the coalition would have been approved and the concerns of those who rely on its services—those on the lowest rung of the stratification ladder—would have been ignored. Fortunately, "in 2009, the City and County appropriated more than $5 million to build a new and improved facility for the homeless—on the current site."

Jonathan White ends this chapter with the piece "Sociology *Is* Action: Using Sociology for Children's Rights," his moving explanation for why he is a Sociologist in Action. In doing so, he describes his work with Free The Children (FTC), an organization dedicated to empowering youth to help exploited children across the globe. White's sociological expertise has helped FTC grow from an organization composed of a small group of students in a single school to an international organization helping hundreds of thousands of children on the lowest rung of the global stratification ladder while "mobilizing more than one million youth to work toward creating the social change they desire." For White, "sociology *is* action," and his work with Free the Children is an incredibly strong example of the potential of sociology to make a positive impact when put into action.

HOUSING AND HOMELESSNESS IN MAINE: A CASE OF PUBLIC SOCIOLOGY IN PRACTICE

Joe Bandy

Vanderbilt University, Nashville, Tennessee

Craig McEwen

Bowdoin College, Brunswick, Maine

Joe Bandy was an associate professor of sociology at Bowdoin College from 1998 until 2010 and is currently the assistant director of Vanderbilt University's Center for Teaching. His research and teaching have focused on topics that include economic development in Latin America, environmental justice and labor movements, transnational movement coalitions, as well as class inequality in the United States. He and his students have partnered with many community organizations to research social problems related to poverty and development.

Craig McEwen is a professor of sociology at Bowdoin College where he has taught since 1975. Over his career, he has drawn on his research on juvenile corrections and on dispute resolution to speak for changes in prison policy in Maine and to assist courts in designing and assessing mediation programs. In recent years, he has partnered with community organizations to enable his students to put their skills to use in addressing local issues through research.

In a declaration of national housing policy in the 1949 Housing Act, the U.S. Congress established "the goal of a decent home and a suitable living environment for every American family."[1] Over 60 years later, this goal eludes many in the United States. A third of all Americans do not own their homes, including over half of all African Americans and Latinos, largely because of the lack of affordable housing. Between 2.2 million and 3.5 million individuals, almost 40% of whom are children, were counted as homeless in 2008.[2] The shortage of affordable housing not only helps to create homelessness, but it also drives greater transience (moving from one place to another), which in turn leads to social stigma; stress; poorer physical health; and weaker and less stable ties to family, employment, and education.

Two-thirds of Americans do own their homes, the single largest and, until recently, the most secure investment for households. Yet even for many within this majority, housing is problematic. Increasing housing costs, debt accumulation, and declining real wages in recent decades have made home ownership less secure. More than one-third of owner and renter households spend over 30% of their income on housing costs, which cuts into the resources they could use for other basic necessities, such as food, transportation, child care, and health care.[3] These problems have been compounded by the subprime mortgage crisis of 2008 and 2009.

These problems exist throughout the United States, and Maine is no exception. From 2000 to 2008, median home purchase prices in Maine rose by 62% and average rents by 31%, while median household income

---

[1]U.S. Code Title 42, Chapter 8a, Subchapter I, § 1441.

[2]National Coalition for the Homeless, "How Many People Experience Homelessness?" (NCH Fact Sheet #2, June, 2008). Available at http://www.nationalhomeless.org/factsheets/How_Many.pdf

[3]The Joint Center for Housing Studies of Harvard University, "America's Rental Housing—The Key to a Balanced National Policy" (pp. 16–17). Available at http://www.jchs.harvard.edu/publications/rental/rh08_americas_rental_housing/index.html

increased by only 22%.[4] The gap between housing costs and incomes widens further as a result of relatively high living expenses in Maine, due in large part to transportation and heating costs. Heating oil, in particular, adds significantly to the cost burden in the sixth-oldest housing stock in the United States,[5] where just over 51% of multifamily rental units are over 50 years old, and most homes are heated with oil.[6] Besides creating general housing insecurity, these factors contribute to significant homelessness throughout the state, causing many homeless shelters to overflow regularly and to turn individuals and families in need away. Unfortunately, because the rural homeless—estimated to be 9% of the total homeless population of the United States and roughly 33% of the homeless population in Maine[7]—are largely invisible to the public, too little information exists on the particular problems they suffer and their needs.

It is in this context that, through our courses at Bowdoin College, we have forged research partnerships with local agencies that are working to end homelessness and to provide affordable housing. These courses include Craig McEwen's Maine Social Research seminar, a course dedicated to engaging advanced undergraduates in practical research problems in and around Maine, and two courses led by Joe Bandy, primarily Class, Labor, and Power, a survey course on class inequality and poverty in the United States that is focused around community research projects. The courses also reflect a growing need for independent research about poverty and housing issues for local agencies and advocacy groups, which have vital research needs but not enough staff and funds to address those needs on their own.

Students in Craig McEwen's Maine Social Research course have partnered with the Brunswick Housing Authority (BHA) on several projects connected to housing affordability. In this seminar course, students learn qualitative research methods, read and discuss empirical research and policy debates, and deepen their grounding in sociological understandings of social inequal-

---

[4]Maine State Housing Authority, "They Can Save Your Life, But They Can't Afford to Be Your Neighbors: A Report on Housing Costs in Maine—2008." Available at http://www .mainehousing.org/Documents/HousingReports/HousingCostsInMaine.pdf

[5]Maine State Housing Authority, "Maine Housing Commits More Funds to Make Homes Safe From Lead Paint Hazards" (November 6, 2008). Available at http://www.mainehousing.org/ News.aspx?PageCMD=NewsByID&NewsID=286

[6]Maine State Senate, Housing and Energy Subcommittee. 2009. "Weatherize all Single and Multifamily Dwellings in Maine" (February 2, 2009). Available at http://www.mainehousing .org/Documents/Reports/REPORT-WeatherizeAllHomes10Years.pdf

[7]National Alliance to End Homelessness, "Rural Homelessness." Available at http://www .endhomelessness.org/section/issues/rural. Maine estimates based on data available from Maine State Housing Authority.

ity. Their central responsibility is to design and carry out field research with community partners and to provide reports that inform audiences that include academics, policy makers, social service providers, and the general public. Through their field research, students deepen their research and analytical skills and reinforce their theoretical understandings of social inequality by connecting to the lived experiences and circumstances of their interviewees.

One of the research projects in the Maine Social Research course focused on the federal Section 8 (Housing Choice Voucher) program that provides rent subsidies for eligible individuals and families (generally with incomes that are 50% or less of the area's median income). These vouchers enable households to rent in the private market and to pay no more than 30% of their income for housing costs. Housing Choice Vouchers constitute the major federal initiative to provide housing assistance to lower-income individuals and families. However, across the nation, there are far fewer vouchers than there is need, and less than 25% of those eligible receive them.[8] The reality of long waits for vouchers led to the research question posed by the director of the BHA: "What do families on the waiting list do for housing while waiting for Section 8 vouchers?" By locating and interviewing waiting list families, students learned that over half of those interviewed had been unofficially "homeless" at some point while on the list—living in tents, partially heated garages, or unfinished basements; or crowded into the dwellings of family or friends; or moving from place to place to avoid wearing out their welcome as "guests." Another quarter lived or had lived in unsafe or unhealthy housing—a condemned trailer, poorly insulated apartments, or moldy dwellings that exacerbated asthma. Most of the remaining families spent more than they could afford on rent and had to skimp on other necessities as a result. The individual stories of families provided compelling portraits of unmet housing needs.

Much to our surprise, we learned that no other research appeared to have been done about how families waiting for vouchers manage their housing needs. As a result, this research provides an invaluable service. According to Greg Payne, coordinator of the Maine Affordable Housing Coalition (MAHC), the student research "focuses on questions that have received surprisingly little attention nationwide and by providing these stories and faces, we may be able to change and strengthen advocacy for increasing the number of Section 8 vouchers authorized by Congress" (personal communication, November 25, 2009).

---

[8]Douglas Rice and Barbara Sard, "Decade of Neglect Has Weakened Federal Low-Income Housing Programs: New Resources Required to Meet Growing Need" (Center on Budget and Policy Priorities, February 24, 2009). Available at http://www.cbpp.org/files/2-24-09hous.pdf

The Maine Affordable Housing Coalition (MAHC) conveyed a brief student report of this research to the Maine Congressional Delegation as part of the coalition's advocacy for support of the Section 8 Voucher Reauthorization Act. The National Low Income Housing Coalition posted the research report on its Web site, and excerpts were included in a statistical report on Section 8 vouchers in Maine prepared for MAHC by the Maine Center for Economic Policy. One of the students who worked on the data from this project observed,

> This research showed me that my sociology classes had real-world applications and that research could have a positive impact on social change and policy. It helped me understand the social issues that are often absent from academia and from the political world. As opposed to simply reading studies and journal articles in class, doing sociological research was exciting and helped enrich my college experience by encouraging me to take on my own sociological research on a policy issue related to low-income housing options.

Similarly, Joe Bandy's course, Class, Labor and Power, has offered groups of students opportunities to participate in ongoing research projects on homelessness in the region. It introduces students to formal sociological research on poverty, inequality, and class structure in the United States, including both structural and experiential dimensions. The centerpiece of the course is a group research project conducted with a community partner that engages students in the skills of original sociological research and public engagement. Over two years, seven separate student groups have participated in four ongoing research projects with Tedford Housing and the United Way of MidCoast Maine. Among these projects, two have been particularly educational for community partners and students.

One of the projects involved a total of 13 extended interviews with current and past guests of the Tedford shelters as a way of understanding Mainers' unique experiences of homelessness and how Tedford's services might become more effective. Students reviewed existing research on experiences of homelessness, learned basic interview methods and ethics, analyzed their findings, and then presented the results to their peers and partners. Their findings affirmed existing research[9] regarding the many interwoven causes of homelessness, including limited educational opportunities and attainment, childhood poverty, high costs of living (particularly

---

[9]For example, see James D. Wright, Beth A. Rubin, and Joel A. Devine, *Beside the Golden Door: Policy, Politics, and the Homeless* (New York, NY: Aldine de Gruyter, 1998).

health care expenses), mental health issues, and—most of all—the limited stock of affordable housing. This confirmed Tedford Housing's experiences and enabled them to move more confidently toward efforts to develop affordable housing. The research also revealed the many subtle ways the homeless can feel belittled or stigmatized by the public, confirming the need for Tedford's to implement humane and compassionate case management.

A second project, also conducted for Tedford Housing, was an opinion poll on homelessness in Midcoast Maine. Tedford Housing expressed a need to know more about the ways the general public understood the causes and effects of homelessness, and thus how they may react to the homeless and Tedford's efforts to build affordable housing units in the area. Over the course of two years, two different groups of students administered an opinion survey to a total of 140 residents of Midcoast Maine. The results were mixed. On the one hand, a majority of respondents affirmed some commonly held stereotypes of the homeless (namely, that they are prone to criminality, substance abuse, and mental illness), and nearly 50% did not want homeless shelters or affordable housing near their residence. On the other hand, most respondents believed that affordable housing was in short supply, and that local nonprofits that try to help the homeless should receive more support. The survey data enabled Tedford Housing to shape its endeavors to develop affordable housing in ways that would mitigate public misconceptions and fears associated with homelessness.

In both projects, students had many opportunities to connect their course material to their research, deepening their understanding of both. The projects also allowed students to utilize interdisciplinary problem-solving skills and spend time reflecting on their own class identities and prejudices. In the words of one student, "The eradication of homelessness is a topic I have become increasingly interested in, and this service project verified that this was something I would like to continue working with in the future. The project showed me how urgent the needs were and how I had to do something to help."

Together, the research collaborations in these courses represent local efforts to educate students about fundamental national and regional issues of poverty and inequality, while helping service providers and policy makers move toward their goals of assisting and advocating for those in need. If community organizations, government, and educational institutions can better understand and address the problems of limited affordable housing, one small step may be made toward realizing Congress's 1949 aspiration for a "decent home and a suitable living environment for every American family."

RELOCATING THE HOMELESS—OR NOT!

James D. Wright

University of Central Florida, Orlando

James D. Wright is an author, educator, and the Provost's Distinguished Research Professor in the Department of Sociology at the University of Central Florida. Wright serves as director of the UCF Institute for Social and Behavioral Sciences, as editor-in-chief of *Social Science Research*, and as editor-in-chief of the *International Encyclopedia of the Social and Behavioral Sciences* (2nd ed.), (forthcoming in 2015). He has published twenty-one books and more than 300 journal articles, chapters, and essays on topics ranging from poverty to homelessness to NASCAR. He also serves on the boards of directors of the Coalition for the Homeless of Central Florida and the Homeless Services Network.

Parramore is the historically African American neighborhood of Orlando, Florida, and has long been regarded by downtown business, development, and political interests as the last remaining chunk of redevelopable real estate in the Orlando Central Business District (CBD). As a matter of fact, a great deal of redevelopment has already occurred there. Parramore is home to the Amway Arena (where the Orlando Magic play basketball), a new events center (where the Magic will play basketball starting in 2011), a new federal court house, the new Florida A&M University (FAMU) Law School, an Orlando Police Department facility, and new national offices for Hughes Tool Corporation (the latter also featuring several retail outlets and some high-end condos).

Parramore is also home to numerous social service agencies, including most of the city's homeless shelters and soup kitchens. The largest of these, occupying more than three acres of land, is the Coalition for the Homeless of Central Florida, the region's largest provider of homeless services (with about 700 people per night sheltered at the facility). For the better part of a decade, the announced intention of the city's mayor and the commissioner of the Parramore district has been to relocate the coalition elsewhere and open the land for redevelopment. Indeed, the *Mayor's Parramore Task Force Report* specifically recommends, as a "first step" toward the goal of improving the community's "compatibility" with development goals, "relocating the Coalition for the Homeless and other similar facilities . . . to achieve goals such as neighborhood preservation."[10]

I have served as a member of the coalition's board of directors and as chair of the board's Research and Evaluation Committee for the last seven years. Early in the relocation process, the board developed a set of criteria that would have to be met by any alternative site. These were straightforward and non-negotiable: proximity to critical social services, proximity to bus routes, sufficient land (not less than 5 acres to relieve chronic space and parking shortages), and adequate funding from the city to acquire the site and build a new facility. The city kept proposing sites (all together, six of them over the years) that were flagrantly unsuitable given the stated criteria (in most cases, because the proposed sites were not near enough to downtown services or not adequately served by existing bus routes).[11] When the board and the CEO of the coalition turned down these proposed sites, we were accused of being uncooperative and obstructive. We were standing squarely in the path of progress!

In 2004, my graduate students and I initiated a series of sociological studies to bring the voices of homeless people themselves into this debate. What would it mean to the day-to-day existence of the coalition's clients if the homeless center were moved out of downtown? About half the people served by the coalition are women and their children. How would the children's lives be impacted? In May, we conducted some focus groups with five homeless women and nine homeless men to broach these very topics and establish a baseline of information. Thereafter, each time a new site was proposed, we would load a group of clients into a van, visit the site, then come back and do another focus group. Between 2004 and 2008, about a half dozen of these ministudies were conducted.

Our initial report on this work was entitled *Coalition Residents Speak Out on Relocation of Coalition Services*. The discussion centered on issues of time, inconvenience, cost, safety, relative isolation, and the psychological stressors associated with each of these. Some of the things we learned were predictable and had gone into the original development of the board's relocation criteria, but the homeless participants in our studies also grasped subtleties that had eluded everyone else. Since the current location is within

---

[10]Passage from p. 25 of the *Mayor's Parramore Task Force Report*, accessible at: http://www.cityoforlando.net/elected/parramore/pdf/20040623_ParramoreReport.pdf

[11]All told, eight alternate sites were proposed between 2004 and 2009. Two of them were deemed perfectly suitable by the coalition and the board but were taken off the table as soon as their suitability to the coalition became known. In both these cases, it was political opposition by the Parramore City Commissioner that caused the site to be withdrawn from consideration.

walking distance of the main bus terminal, coalition residents can get pretty much anywhere downtown in 30 to 60 minutes. More remote locations, they feared, would cause them to miss busses, miss appointments, be late picking their children up from child care, or be late for work. They would need to take one bus to the main terminal, then transfer to another bus to get where they needed to go; in one woman's case, moving away from the central location would have added another hour of travel time to and from work and would require her to transfer her daughter to a new school. (This was a problem for several women.) The added travel time would have also made her unable to be home with her daughter before and after school, thus imposing significant child care expenses that she could ill afford.

One proposed location would have required daily treks through an area rife with prostitutes and drug dealers. Many clients, male and female, worried about how this might affect their personal safety, but the women frankly voiced another concern: the very real possibility that they would themselves be mistaken for prostitutes. These women already struggle daily with the stigmas of poverty, homelessness, and (for the majority) race. Adding the misperception that they were also hookers was more than they could bear.

Despite their misgivings, residents were stoic about and resigned to the larger forces in play. One woman said,

> I know whatever place they intend to move us to would be worse than this place. This area is beginning to clean up a little bit and so now they want to make lots of money off us rather than let us stay here. And, maybe us having helped a little bit to make this section of town a better place than it used to be is part of the problem. But we don't see that credit being given to us. There's 200 of us in this tiny building that want better than what we had when we got here.

And in the same vein, another client stated, "I know without a shadow of doubt that if they move us, it's going to be in a worse area of town. I know they wouldn't find us a better place." With the two noted exceptions, the proposed sites confirmed the accuracy of these perceptions: an abandoned industrial site with no nearby amenities or services, a drug-infested area far from the city core, a site bordered by active and busy railroad tracks, and so on.

At about the same time, we opened a separate line of research investigating the experiences of other cities that had attempted to relocate their homeless services outside the CBD. Orlando was certainly not the only city with economically depressed near-downtown areas housing lots of homeless facilities, nor the first to imagine moving those facilities elsewhere. My contacts in the network of homeless researchers quickly yielded a sample of cities that had tried to do this (Chicago, New York City, Tallahassee,

Tucson, Columbus, San Diego, Los Angeles, Charlotte, Atlanta, and Ann Arbor), and a few e-mails and phone calls brought the information we sought. Most cities that had attempted to relocate homeless services away from their CBDs, we learned, had been prevented from doing so by a combination of costs, NIMBYism, and local politics.[12] Moreover, the few cities that were successful in their relocation efforts did *not* experience much (if any) downturn in the numbers of vagrants, panhandlers, or homeless people hanging around in the CBD. The universally hoped-for decline in the presence of apparently homeless persons in the CBD, in short, never materialized. My correspondent at the National Coalition for the Homeless (the nation's leading advocacy organization for the homeless) explained,

> It is my opinion that a certain segment of the homeless population will go to wherever the shelter is, as they have no other choices. This is especially true for families, the elderly, and people who are somewhat institutionalized to the shelter system/life. But a large segment of the homeless population (especially singles) will continue to hang out in the downtown areas. After all, this is their home.

In another investigation, we surveyed the people in line at the coalition's evening feeding program. We learned that a fifth of those in line were *not* homeless people, but rather poor people and families from the surrounding neighborhoods who were using the feeding program to help stretch their budgets and make ends meet. If the coalition were relocated to a remote site, where would these hungry individuals and families eat? When concerns were expressed about the "bums and beggars" that lived at the coalition's Men's Pavilion, we undertook a survey of the men about their economic activities. Two out of three spent their days working or looking for work (often in the nearby day labor outlets); more than half of the remaining third were disabled and literally unable to work. As for panhandling, only about one in six of the men *ever* panhandled, and only about 2% said they did so regularly.

Because of my position on the coalition's board of directors, it was an easy matter to feed our data, findings, and interpretations to the board leadership and to our Government Relations Committee, which in turn fed what we had done into the larger political debate. None of the research findings was itself *decisive*, but we did give voice to people who were other-

---

[12]NIMBY stands for "Not In My Back Yard" and indicates the mind-set of communities who are fighting to keep something they view as detrimental to their area (e.g., a big-box store like Wal-Mart, a prison, low-income housing, a halfway house for sexual offenders, a nuclear plant, etc.) out of their local neighborhood. Communities everywhere view homeless shelters as detrimental, which is why they are frequently located in neighborhoods whose residents lack political clout (e.g., high-poverty or historically African American communities).

wise voiceless. In nontrivial ways, we enabled their participation in a debate whose outcome would directly affect them and their lives.

In 2008, the relocation issue was finally resolved: The Coalition for the Homeless would be allowed to stay where it was indefinitely. In 2009, the city and county appropriated more than $5 million to build a new and improved facility for the homeless—on the current site.

A Los Angeles advocate, commenting on that city's efforts to relocate homeless facilities, made a compelling sociological observation:

> I think the most damaging aspect of the relocation proposal is that it assumes an otherness about homeless people that will only serve to make it more difficult for agencies to help them in the community at large. Are they or are they not part of your community? If they are not, then a policy of ostracizing them fits (however morally reprehensible). If they are, then the community leaders ought to take another look at this proposal and—dare I say it?—show some leadership in the face of likely vocal opposition.

Or, as our Atlanta contact expressed it,

> The trick for all of us who are involved in this work is to educate folks to consider the possibility that homelessness is an experience, not a blood type. People who experience the lack of housing, employment, treatment, etc., need those services and must be reintegrated into our communities.

This is reminiscent of a prescient remark by a homeless woman in Kim Hopper's excellent ethnography, *Reckoning With Homelessness:* "You have to understand that this is a *condition,* this homelessness. It's not who we *are*" (p. 142, emphasis original).

Postscript by James Wright: Although several students developed conference presentations out of the body of work summarized above, none of it has ever appeared in a refereed journal article. But it was some of the most rewarding and intellectually satisfying work of my 35-year career.

Postscript by Dr. Amy Donley, lead graduate student on the coalition work: I first went to the coalition as a novice master's student to gain some research experience. Little did I know that my intellectual path would be transformed. I have since worked on virtually every homeless project in Jim's shop: a study of homeless people living in the woods of East Orange County, the annual point-in-time homeless counts, a study of the elderly homeless (which has just been accepted for publication), and a needs assessment survey of homeless men to assist in the design of the new shelter that is now to be built on the coalition site. My dissertation research focused on perceptions of the homeless as dangerous. What began for me as a volunteer project has become my passion and life's work. After that first night at the coalition, I drove home in tears—partly tears of pity for the wretchedness I had witnessed but also tears of rage that I lived in a nation that would allow something like this to happen.

Reference

Hopper, K. (2003). *Reckoning with homelessness*. Ithaca, NY: Cornell University Press.

SOCIOLOGY *IS* ACTION: USING SOCIOLOGY FOR CHILDREN'S RIGHTS

Jonathan White

Bentley University, Waltham, Massachusetts

Jonathan White studies and teaches about inequality, globalization, human rights, and public sociology and is the Director of the Bentley Service Learning Center at Bentley University. He is coauthor of *The Engaged Sociologist* (2012, with Kathleen Odell Korgen), served as associate editor to the *New York Times* best seller *Me to We* by Marc and Craig Kielburger (2008), and is currently writing a book on hunger in the United States. Dr. White has received teaching awards at Bridgewater State University, Framingham State University, Colby College, and Bowdoin College. He is the founder of Sports for Hunger and the Hunger Resource Center, and serves on the board of directors for Free the Children, Peace Through Youth, and the Graduation Pledge Alliance.

A public sociologist is the only type of sociologist I know how to be. I was "raised" as a sociologist through the undergraduate program at Brandeis University and the graduate program at Boston College. In both of these programs, I learned that being a good sociologist means bringing our skills, training, and energies to working for the public good. I have always been most interested in social stratification—in the ways that members of societies are ranked in systems that allow some to "win" and some to "lose." And I am particularly compelled to learn about those who find themselves on the lower rungs of stratification, facing lives of desperate poverty.

In 1998, I was researching the issue of child labor for a lecture I was preparing for one of my courses. The scope of this issue, one of the worst social problems stemming from a global system of stratification, is staggering. An estimated 225 million children work illegally worldwide, and nearly half of these in the most dangerous and appalling labor industries, such as sex work, child soldiering, bonded labor, and slavery. Behind the statistics and studies, I was also reading about the compelling human stories of pain and suffering that permeate the lives of these children and their families.

Enmeshed in (and depressed by) my research, I came across an inspiring story of a 12-year-old boy from Canada, Craig Kielburger, who, like me, had been learning about child labor but who had decided that he needed to do something to help! Mustering up the courage to speak in front of his schoolmates, Craig taught them what he had learned about child labor and asked if some of them would join him to figure out how to help. Several volunteered. They began educating others about child labor and raising money to help. Compelled to learn more so he could make an even bigger difference, 12-year-old Craig gained his parents' consent to travel to southeast Asia (with an older friend acting as his guide) to meet children his age who were child laborers. While there, Craig met kids whose lives at the bottom of the global stratification system were horrifically different from his back home in Toronto. Moved by their stories, he successfully arranged a meeting with then–Canadian Prime Minister Jean Chretien and pushed for a "no child labor" clause to be included in the Canadian–Pakistani trade agreement. Upon Craig's return to Canada, the media picked up the story and spotlighted the work of Free The Children (FTC), a group Craig cofounded with his older brother, Marc. Inspired even more to make a difference, Craig worked tirelessly with Marc and the other youth members of FTC, and soon their organization began to expand enormously.

Reading this incredible story about the efforts of FTC's youth members, amidst all of the stories of suffering and abuse of child laborers, I felt inspired and invited Craig and Marc to speak to my students at Framingham State College (now University), where I was teaching at the time. Their talk that evening, one I will never forget, interweaved painful stories of child laborers they had met on their journeys with heart-inspiring stories of young people in North America working with FTC to try to help. Later that evening, Marc, Craig, and I spent hours sharing information and strategizing about how to further organize for children's rights. I began working with FTC right away and have continued ever since. I saw clear ways I could use my sociological training and research in human rights issues to help as FTC grew.

That summer, I traveled to Kenya with FTC's youth leadership partner organization, Me to We, with Craig, Marc, and a group of other FTC youth. We spent our mornings learning about local culture and daily life from the people in nearby towns, our afternoons teaching leadership skills to the youth members, and our evenings by the campfire where I delivered talks to contextualize the poverty and social problems we were witnessing.

That summer, I also spent a week at Free The Children's Take Action Academy delivering workshops to over 100 of the top FTC youth leaders from around North America and Europe. Over the past decade, I have continued to take advantage of the opportunity to deliver talks and workshops to FTC youth and to work with FTC leadership, including Craig.

My goal is always to help the staff and youth membership to gain a deeper sociological understanding of the social forces promoting inequality and stratification that work to deny children their human rights.

In 2000, the United Nations approached Free The Children to become a lead partner in the upcoming Decade for Children and Armed Conflict. FTC asked me to help set up a large-scale global project enabling youth to learn about and address issues related to children and armed conflict. I set up a team of nearly a dozen of my students (the majority of whom were sociology majors) who volunteered to assist with the research and programmatic aspects of the project. Working alongside the talented staff in the FTC office, we created the Youth Ambassadors for Peace Project, mobilizing youth from nonconflict regions to raise awareness about the issues of children and armed conflict and raise funds to help youth in postconflict and current conflict regions. We researched issues, produced publications, created campaigns, and worked with an active and passionate FTC Youth Board. War Is Not a Game was one fundraising campaign whereby youth from non-conflict zones turned in their war toys, making a collective statement that "while we play war, for millions of children around the world, war is not a game." The Schools for Peace campaign mobilized youth in nonconflict regions to raise money to build schools in postconflict regions, such as Sierra Leone, to help communities begin to rebuild after the devastation of war. For me, certainly, one of the most satisfying and powerful experiences of this project was watching my students become skilled sociologists in action!

In 2005, Craig was engrossed in his studies at the University of Toronto and had become interested in learning more about the socio-psychological processes that lead to *bystanding* (when people stand by and allow injustices to continue rather than standing up to eliminate those injustices). He asked me to conduct a guided independent study course with him on the topic. It was an intense semester of reading and researching for both of us, and, as with much good research, it led us to more questions than answers. Ultimately, Craig worked toward a deeper understanding of why more ostensibly good and moral people don't take action to create a more socially just world.

The guided research course helped lay a foundation for Craig and Marc's book, *Me to We: Finding Meaning in a Material World* (2008), which explores bystanding, points to developing empathy and acts upon this empathy as an antidote to bystanding, and provides motivational stories and tools illustrating how we can become more active and engaged global citizens. I had the privilege of serving as an editor of the book, which became a *New York Times* best seller and jump-started a Me to We social movement—a movement that, in turn, has engaged hundreds of thousands of students, educators, and other concerned citizens interested in moving from the "me-centered," individualistic philosophy that permeates our culture toward creating a more "we-centered" society.

> Through the writing stages of our book, *Me to We: Finding Meaning in the Material World,* we traveled to Boston to learn from Dr. White. Through his mentorship, he helped us develop our understanding of the bystander effect, the role of empathy as a guiding force, and the sociological research behind these crucial concepts. Our understanding of these social phenomena went beyond our book and helped to inspire a "Me to We" movement . . . helping to engage tens of thousands of young people to develop their empathetic understandings of their role as global citizens.
>
> —Craig Kielburger

As Free The Children developed from a small nongovernmental organization (NGO) operating out of the Kielburger family home into a much larger organization, there have been many ways I have been able to use my sociological skills to assist the organization. One big question, for instance, has been how to seek corporate donations while assuring they do not come from companies that use child labor or are otherwise connected to children's rights abuses. My sociological research into fair trade, labor standards, human rights, and labor monitoring has helped the FTC development team devise a corporate vetting system.

> As we move forward developing both Free the Children and Me to We, Dr. White's expertise and insight into human rights, corporate social responsibility, and social entrepreneurship have guided us through every phase of our work. His advice has helped us remain true to our mission and vision as we expand our funding partnerships.
>
> —Marc Kielburger

I have also helped FTC create several additional youth-empowering, sociologically informed campaigns. One of my favorites is Halloween for Hunger, which I originally created with a group of my college buddies and later developed more fully while conducting research on U.S. hunger for my dissertation. I interviewed dozens of the more than 50 million hungry Americans and conducted a survey of nonhungry college students to determine what they knew (and didn't know) about this issue. While conducting my research, I founded a nonprofit organization, Sports for Hunger, to raise awareness and funds for the alleviation of U.S. hunger. One of our key campaigns was Halloween for Hunger, which, as Craig notes below, later became a major campaign for Free the Children.

> One of our most successful campaigns is Halloween for Hunger, an initiative created by Dr. White. Our youth members mobilize by raising awareness and funds for hungry individuals in their region, as they trick-or-treat on Halloween for canned goods for those who are hungry, instead of candy for themselves.
>
> —Craig Kielburger

Over the years, I have also helped Free the Children develop other similarly sociologically informed campaigns, such as Brick by Brick (in which youth raise funds to build schools in poor countries where children otherwise might not have an education), Celebrate for Change (where youth use their birthdays to raise funds to help disenfranchised youth around the world), the Vow of Silence (a 24-hour campaign whereby youth seek sponsorship for each hour they remain silent), and 10-by-10 (young people commit to 10 social change actions in exchange for $10 from a foundation partner, which is then used to fund FTC's amazing development projects). It is my hope that my sociological expertise can help participants to both gain a deeper understanding of the extreme difficulties many children face *and* to act in solidarity with these children to address global economic inequities.

During the past 13 years, I have also been honored to serve as a member of Free The Children's board of directors.

> Since the early days of Free The Children, Professor White's presence has been a constant. His sociological expertise has made a mark on every one of our projects and programs. Most importantly, his mentoring role and presence as a sociologist and expert on human rights issues has allowed us to grow on a personal, professional, and organizational level.
>
> —Marc Kielburger

Since happening upon that article in 1998, I have watched FTC grow into a powerful organization, mobilizing more than one million youth to work toward creating the social change they desire. To date, FTC has built over 650 schools in some of the world's poorest regions; delivered over $16 million worth of medical supplies; provided access to clean water, sanitation, and health care to over one million people; enabled over 30,000 women to become involved in microdevelopment projects; and empowered tens of thousands of North American youth with the leadership training and skills they need to be effective social change-makers. The fact that *youth* members have raised more than two-thirds of the money contributed to FTC's projects is just one indication of FTC's successful work in empowering youth to work toward the social change they desire.

FTC has received great recognition, including the 2006 World Children's Prize for the Rights of the Child, also known as the Children's Nobel Peace Prize; the 2006 WANGO Human Rights Award from the United Nations; the 2007 Skoll Award for Social Entrepreneurship; the World Economic Forum Medal; the State of the World Forum Award; the Roosevelt Freedom Medal; and three nominations for the Nobel Peace Prize. Each year, FTC kicks off annual campaigns with our "We Days," where tens of thousands of youth members come together in stadiums for a day of inspiration and action. Broadcast live by MTV Canada, these events reflect FTC's remarkable support, with attendance by Nobel Peace Prize winners, such as Elie Wiesel, Jane Goodall, Desmond Tutu, the Dalai Lama, Betty Williams, and Wangari Maathai, as well as celebrities and activists, such as Mia Farrow, Demi Lovato, Magic Johnson, the cast of Degrassi High School, Sarah McLachlan, Robert Kennedy Jr., Romeo Dallaire, Justin Bieber, Martin Sheen, K'Naan, the Jonas Brothers, Jason Mraz, and many others.

My personal rewards for my work with the Free The Children and Me to We organizations have been immeasurable and deeply humbling, and doing this work is what makes me truly proud to be a sociologist. For me, sociology *is* action. It is an opportunity to live Mahatma Gandhi's plea that "Each of us must be the change we want to see in the world."

Reference

Kielburger, C., & Kielburger, M. (2008). *Me to we*. New York, NY: Fireside Press.

DISCUSSION QUESTIONS

1. Bandy and McEwen point out that they were surprised that no one before them had conducted a study on how families waiting for vouchers meet their housing needs. Why might this population be relatively ignored by the general public, policy makers, and researchers? Why do you think sociologists should study this social situation?

2. How did James Wright and his students give "voice to the marginalized"? Think of a marginalized group on your campus or that you have observed in your local community. Describe them and what makes them marginalized in their community, and then write about how you might create a study that helps to "give voice" to their experiences.

3. Wright's contact from Atlanta said, "The trick for all of us who are involved in this work is to educate folks to consider the possibility that homelessness is an experience, not a blood type." What do you think he means by this? How do you view homelessness? Why do you see it that way?

4. Jonathan White describes many sociologically informed campaigns that enable youth to help those in need. If you were asked to choose one of these campaigns to conduct, which would it be? Why? How would you go about carrying out the campaign?

5. Why has White's sociological perspective been so important to the success of Free The Children? Why is it important to understand a society's system of social stratification in order to effectively address inequality?

6. Use at least two of the Sociologist in Action pieces in this chapter to provide examples of how sociology can be used to empower members of society on the lowest rungs of the social stratification ladder. After you provide your examples, describe what might have happened if the Sociologists in Action you discuss did not carry out the work you described in your answer above.

7. How do you think your own social class impacted your feelings when reading the Sociologist in Action pieces in this chapter? Why? How does our social class shape our perspectives on members of both our own and other social classes? If you were to become a Sociologist in Action, how would you keep aware of your own social class and attempt to keep it from biasing your research and work?

RESOURCES

The following Web sites will help you to further explore the topics discussed in this chapter:

"People Like Us: Social Class in America"	http://www.pbs.org/peoplelikeus/
Class Matters	http://www.nytimes.com/pages/national/class/index.html?8qa
Gapminder	http://www.gapminder.org/
LIS	http://www.lisdatacenter.org/
Sociosite Stratification	http://www.sociosite.net/topics/inequality.php
U.S. Census Bureau: Housing Topics	http://www.census.gov/housing/

To find more resources on the topics covered in this chapter, please go to the Sociologists in Action Web site at **www.sagepub.com/korgensia2e.**

Chapter 9

Race and Ethnic Relations

R ace and ethnic relations have been a focus of sociology ever since
W. E. B. DuBois was on the scene over 100 years ago. However, since
DuBois' day, sociologists have not always been at the forefront of those
confronting racial injustice (Steinberg, 2007). Many modern public sociologists, though, are following in the footsteps of sociological founders
like DuBois, Jane Addams, and George Herbert Mead and putting their
sociological knowledge to use in the effort to promote racial and ethnic
equality in the United States. The Sociologists in Action featured in this
chapter provide great examples of how sociological tools can be used to
promote racial justice.

In "Bridging the Campus and the Community: Blogging About the Asian
American Experience," C. N. Le shares how he began to learn about and
embrace his Vietnamese and Asian American heritage after he began to take
some sociology courses in college. He decided to use his sociological knowledge to embrace the expectation that he would speak up for other Asian
Americans. Through establishing his Asian-nation.org Web site and blog,
Le has "portray[ed] Asian Americans as accurately and comprehensively as
possible, rather than let[ting] . . . other Americans rely on distorted portrayals and ignorant stereotypes about Asian Americans." In the mode of public
sociology, he has made sure his postings are as jargon free as possible in order
to bridge the campus and community divide. Le points out that, in addition
to educating the public at large, his Web site is "a source of information and
learning for young Asian Americans, many of whom grow up [as Le did]
isolated from their history, culture, and collective experiences." His Web site
and blog are also a means "to mobilize [the Asian American] community in
times of crisis (e.g., responding to a high-profile incident of racism)." Both

are excellent examples of the use of sociological tools to combat stereotypes, prejudice, and discrimination.

Barbara Gurr also describes how to alleviate racial discrimination in "The Responsibilities of Relationships: Using Sociology to Build Meaningful Alliances." In this piece, she relates her efforts to use her sociological tools to build alliance structures between Native and non-Native Americans. Through her institutional ethnography of Native American women's health-care, Gurr learned that "reproductive healthcare for Native women is deeply influenced by the dominant culture's ideas about race, gender, sexuality, and citizenship." Gurr uses her findings and sociological imagination to advocate for better healthcare for Native women and to create and strengthen alliances between Native and non-Native peoples.

In the third piece in this chapter, "Putting Sociology to Work in Winnersville USA.," Mark Patrick George describes how he uses sociological tools to fight racial injustice in the educational system in his hometown of Valdosta, Georgia. He has used sociological data to "out" the fact that over six decades after segregated schools were deemed unconstitutional, the Valdosta county schools (VCS) "remain a system controlled by white administrators and educators . . . that 'works' primarily for its white minority population." He shows that his efforts "demonstrate the usefulness and power of social research and asking critical sociological questions." George's work has also led to tangible social changes that have sparked a renewed community drive to make the VCS work for students of all races.

David Cunningham closes this chapter with "Methods of Truth and Reconciliation," in which he describes how his applied research methods class at Brandeis University partnered with the Mississippi Truth Project to help them "connect the dots between past discrimination and racial inequity in present-day Mississippi . . . [and] to 'shape an inclusive and equitable future.'" This partnership enabled students to develop their research skills while working for social justice. They analyzed the connection between the racially discriminatory practices of various institutions in Mississippi before the civil rights laws of the 1960s were enacted and racial inequality today. Cunningham concludes by noting that, "as a sociologist, there is nothing more satisfying than linking students' passion for social justice to the tools of social science research, and witnessing the possibilities for change that result."

Reference

Steinberg, S. (2007). *Race relations: A critique.* Palo Alto, CA: Stanford University Press.

BRIDGING THE CAMPUS AND THE COMMUNITY: BLOGGING ABOUT THE ASIAN AMERICAN EXPERIENCE

C.N. Le

University of Massachusetts Amherst

C.N. Le is a senior lecturer in sociology and is director of the Asian/Asian American Studies Certificate Program at the University of Massachusetts Amherst. His work as a Sociologist in Action includes working for the National Asian Pacific American Bar Association, as director of education for the Asian Pacific Islander Coalition on HIV/AIDS, and as a research associate for the Center for Technology in Government at the University at Albany SUNY. He also tries to put his academic, professional, and personal experiences to good use by blogging about issues related to Asian Americans and racial/ethnic relations at his Web site, Asian-Nation.org.

What do you do when you feel like you're the only one around? That's the question I frequently faced growing up in a predominantly white society. I had a lot of friends growing up, but almost always I was the only Vietnamese immigrant, the only Asian American, and the only person of color in my school and my neighborhood. I was expected to "represent" or be the spokesperson for these minority groups in a lot of situations. I didn't want that responsibility. Instead, I wanted to just blend in, to not stand out as being different or strange. In many ways, I wanted to be like my friends—to be white. Through those elementary, middle school, and high-school years, I did not really emphasize my minority identity, opting instead to deemphasize it as much as possible.

It wasn't until my junior year of college when I decided to minor in sociology and began taking classes on race and ethnicity, immigration, and Asian American studies that I finally began to learn about the history of people of color, immigrants, and Asian Americans in the United States. I learned how, in the face of injustices, inequalities, and oppression, people of color have not only survived but become stronger and more resilient through these experiences. I finally began to experience my Vietnamese and Asian heritage as a source of pride and confidence instead of shame and embarrassment. With this newfound understanding of myself and the world around me, I began to claim and embrace the expectation I had once shied away from: the responsibility of speaking up for others like me.

As I progressed through my research on various aspects of Asian American assimilation, the Internet revolution was well underway, and blogs, in particular, were becoming quite popular. I saw this as an opportunity to leverage the emerging power of the Internet by blogging to share what I was learning about the collective experiences of Asian Americans in particular (and immigrants and people of color more generally). It was my hope to portray Asian Americans as accurately and comprehensively as possible, rather than let others portray my community however they want and have other Americans rely on distorted portrayals and ignorant stereotypes about Asian Americans.

With that in mind, I started my Asian-Nation.org Web site and blog in 2001. From the beginning, I wanted my site to be as accessible as possible. On my Web site and through my blog, I make "academic" research and data easy to understand. I also cover as many real-world issues and social problems in American society as possible. You might say that I have been able to bridge the gap that separates traditional academic articles from those published in magazines and newspapers—to connect the campus and the community.

I try to bridge these two aspects of society by first acknowledging that I am not always going to be completely objective or unbiased in my writing. As I tell my students, many academics claim to practice total objectivity in their research and teaching. Other professors think that it is virtually impossible to keep one's personal beliefs completely separate from one's research or teaching, and I happen to agree with the latter group. Biases can take many different forms—some may be quite blatant in terms of direct statements or writings, but other examples can be more subtle, such as when sociologists decide which questions to ask, which methods and data sources they'll use to answer those questions, and how they present their findings. In fact, as we see over and over again, the same statistical data can be used to support completely opposite interpretations and sides of a debate. With that in mind, I am up front with the students in my classes and the readers of my site and blog. I let them know that, based on both my academic training and my personal experiences, the political and social views I have tend to be liberal in nature. However, I also tell them that as a sociologist I am compelled to back up what I say and what I write with established theory, reliable data, and appropriate supporting examples.

I try to achieve a yin–yang balance between objectivity and subjectivity. Being an *academic blogger* means that people may pay a little more attention and perhaps give a little more credence to what I have to say because of my position as a professor. But with that added respect comes added responsibility. In other words, I can't get away with just ranting and raving about racial incidents or other injustices going on around me. Instead,

I understand that as an academic I have to support my opinions and back up my criticisms with valid theories, data, and intelligent analysis.

I think that sociologists are well-qualified to do this kind of work. I firmly believe that sociologists should apply their experiences and their expertise to educate as many people around them as possible (not just other faculty or college students), and to apply their work to inform and influence social issues and policy to make sociology as "public" as possible. Sociologists can apply their research to lend some objectivity and empirical data to often emotional discussions and debates around controversial issues. I understand perfectly well that sound data and research may not ultimately sway people with passionate opinions, but I feel I've done my job if I help people look at issues more comprehensively and from different perspectives.

Beyond using my Web site and blog to educate the general public about the histories, experiences, and characteristics of Asian Americans, immigrants, and people of color, I also hope that my efforts can benefit the Asian American community specifically. That is, I hope that blogs like mine will be a source of information and learning for young Asian Americans, many of whom grow up isolated from their history, culture, and collective experiences, similar to what I went through when I was their age. I felt very satisfied when my site was mentioned in media outlets, such as *USA Today, Yahoo News,* the American Press Institute, the Library of Congress, PBS, and the *Washington Post,* to name a few. However, I am particularly gratified when I receive comments from ordinary readers, such as the following:

Thank you for putting all this research together in one place! For most of my life, I haven't really paid much attention to Asian American issues. In school, I was always doing projects on the plight of the Native Americans, the Holocaust, or slavery, but I was never really encouraged to learn about Asian American History.

I am so grateful that you are doing this. You have no idea how relieved I was to find this site . . . . I always realized the misrepresentation of Asian Americans in this country, but had no idea just how far it extended.

Yours is the best website that represents the Asian American community I have seen so far. It is very well-organized, informative, and most importantly, it has a certain candid and personal tone which I feel makes it very distinctive from other sites.

Web sites and blogs also can help Asian Americans develop networks of people who share similar concerns, experiences, expertise, and so forth that can be used as a source of social support, but also to mobilize our community in times of crisis (e.g., responding to a high-profile incident of racism). A sampling of these responses in recent years includes Asian American blogs like mine joining with national advocacy organizations, such as the Asian American Justice Center, the Asian American Legal Defense and Education Fund, the Media Action Network for Asian Americans, the Japanese American Citizens League, and the Organization of Chinese Americans to protest racist and stereotypic media portrayals of Asian Americans in the 2009 movie *The Goods* and *Details* magazine's "Gay or Asian" column. We also took legal action to protect the rights of Asian American high-school students in Philadelphia, Pennsylvania, who endured a series of physical attacks in 2009, with school officials turning a blind eye to their repeated requests for help. Perhaps most famously, we advocated for nuclear physicist Wen Ho Lee who was falsely accused of espionage by federal authorities in 1999.

I think that Asian American Web sites and blogs like mine demonstrate that we want to be part of the American cultural mainstream and engaged in the communication and democratic processes, rather than secluding ourselves from the rest of American society as critics sometimes charge. Also, Asian Americans can use blogs and other emerging forms of Internet technology, such as Facebook, Twitter, YouTube, and so on to be at the forefront of technology and Internet culture—to be "cultural entrepreneurs," like our technological counterparts who are achieving success in Silicon Valley. Finally, these forms of technological expression give us a voice that we haven't had before—another mechanism to speak up and empower ourselves to overcome the ways in which we've been silenced through the years by the cumulative effects of racial stereotypes and discriminatory laws.

With these points in mind, my site and blog are basically extensions of my research and teaching—they are an integral part of who I am as a scholar and an Asian American. On top of that, the topics that I am most interested in—race relations, immigration, assimilation, politics, and the cultural effects of globalization—are some of the most controversial and important issues facing American society today. Since these issues are still being played out and will continue to evolve for the foreseeable future, I anticipate having plenty of material to write about. For me, being a Sociologist in Action means fully expressing myself academically and personally, educating and empowering others, and contributing what I can to the larger human community.

THE RESPONSIBILITIES OF RELATIONSHIPS: USING SOCIOLOGY TO BUILD MEANINGFUL ALLIANCES

Barbara Gurr

University of Connecticut

Barbara Gurr is an assistant professor in residence with the Women's, Gender, and Sexuality Studies Program at the University of Connecticut. Her research on Native American women's reproductive healthcare has been published in *The International Journal of Sociology of the Family, Sociology Compass, The Journal of the Association for Research on Mothering*, and other locations. She has won numerous awards for her research, teaching, and activism and is inspired and motivated on even the worst Mondays by the generosity and strength of her relatives in Indian Country as well as the generosity and curiosity of her students.

Quick Quiz for the Non-Native American:

1. Name ten Native nations and where their reservation homeland is. If that's too difficult, how about five?

2. Name five health issues facing Native people today.

3. Name one piece of federal legislation that impacts Native people today.

4. Where is the poorest county in the country? The second poorest? The third?

5. Who are your relatives?

If you're like most non-Native people in the United States, you had trouble with this quiz. This is not (entirely) your fault. Unless you live on or near a reservation homeland, are majoring in Native Studies, or hanging out regularly with Native people, your opportunities to learn about these things have been severely restricted by your education. I'm not Native either (by the way, if you're reading this and you *are* Native, welcome, and thank you for your patience as I work through some of this for our relatives). But *alliance*—a slippery concept at times, particularly perhaps across racial lines—is important to me as a feminist sociologist. In fact, most of my work derives from my (sometimes awkward) efforts to build alliance structures between Native America and non-Native America. Sociology has provided me the tools with which to do this. Relationship is also important to me. There is a common phrase in Indian Country, frequently heard in prayer ceremonies and at other important moments: *Mitakuye Oyasin*. It translates to "we are all related" or "all my relations." It has deep significance for the Lakota people, many of whom take its meaning to heart in their everyday lives and actions. I try to do the same. After all, if we are all related, if we are all relatives, then shouldn't we be doing our best to understand each other's lives? Again, sociology has provided me with the tools with which to do this.

But back to the quiz: few people in the United States who are not Native American understand the complexities of Native-U.S. relations. Of course, not all Natives do, either. It really does get complicated, involving Supreme Court decisions; hundreds of treaties, congressional acts, federal, regional, state, and tribal laws; state produced and enforced notions of identity, personal and community notions of identity . . . you get the picture. Complicated. In fact, few people in the U.S. who are not Native even realize there is such a morass of complexities as those covered under the term *Native sovereignty*, although this is an increasingly hot topic in areas where Native nations have recently gained federal recognition, or—horrors!—plan to build a *casino*. Sociology has helped me learn about these things and provided me with the tools to become a better relative of Native people.

I began an intimacy with Indian Country in the summer of 1999, when I traveled out to Pine Ridge Reservation in South Dakota, home of the Oglala Lakota, to build wheelchair ramps. Of course, there are plenty of Native Americans in Connecticut, where I grew up (just as there are plenty in your home state, whatever it is), and I didn't have to go all the way to South Dakota to learn about Native people. But my family actually has a long history of near-misses with Pine Ridge, ancestors and relatives who have wanted to go, planned to go, but somehow never got there. So when I had the chance, I took advantage of it. Building wheelchair ramps was just my way in, but it wound up being a profoundly informative way, introducing me to the problem of diabetes in Indian Country, where diabetes is approximately 500% percent (that's right, I said 500%) higher than in the general population.

When I left Pine Ridge after that first trip, I knew I'd return. A year later, I moved to Pine Ridge to teach English at one of the few high schools on the reservation. That's also the period of time when I began the process of becoming a sociologist, though I didn't officially begin my graduate studies until six years later. I went to graduate school for the specific purpose of studying Native American women's reproductive healthcare, something I became particularly interested in when I lived on the rez.

Healthcare sounds like it might be better understood through a public health, rather than a sociological, perspective. Fortunately, though, I had a graduate advisor who recognized me as a sociologist long before I did. She encouraged me to pursue a PhD in sociology and do what's called an *institutional ethnography* (IE) for my thesis and dissertation. Institutional ethnography was first developed by Dorothy Smith (1987, 2005) as a "sociology for people." Sociologists who use this approach study the institutional organization of our lives, asking questions about the ways in which our experiences are shaped by institutions, such as healthcare and education (instead of the other way around!). Institutional ethnography, and sociology in general, allowed me to see things about Native women's healthcare I might not have seen if I had approached it from any other angle.

In order to do research with a Native community, permission must be sought from the members of that community, and ideally members of the community are involved with all phases of the research. I worked hard to do this, meeting with community members and members of the Tribal Research Review Board regularly to determine what *they* thought needed to be done and how *they* thought it should be done. I took their teachings seriously, trying to design a project that would be meaningful to the communities with which I worked. This was empowering for both me and my Native relatives, as our diverse voices and needs blended together in different ways to produce research and findings that could serve us all.

My research included mapping the multiple authorities that influence the Indian Health Service (a federal agency created to provide healthcare to Native Americans, a right guaranteed through hundreds of treaties, pieces of legislation, and congressional actions), including tribal councils, regional administrators, regional state governments, the Department of Health and Human Services, the president of the United States, and others. This *map*, a common technique in IE, reveals where and how certain decisions about Native healthcare are made, as well as where and how certain alternatives might be available. Primarily, we learned about how complex health and healthcare are; for many people, including many Native Americans, it's not a simple matter of eat right, get plenty of exercise, and go see the doctor when you're sick. The role of the federal government and tribal governments in organizing reproductive healthcare for Native women is deeply influenced by the dominant culture's ideas about race, gender, sexuality, and citizenship. This results in, among other things, gross underfunding of Native healthcare, which means outdated facilities, inadequate staffing, and limited services. For Native women, it often means severely restricted access to contraception, prenatal care, and even mammograms—health care I and most of my students generally take for granted.

Today, I continue to work closely with tribal members, tribal and pan-tribal organizations, and healthcare providers and advocates who work for and with the Indian Health Service. Sociology has provided us with ways to understand Native women's reproductive healthcare as delivered by the "State" (the same loosely organized affiliation of ruling apparatuses which quite possibly produced, and at the very least deeply influenced, your education—and the gaps in your education. Need I refer you to the quiz again?). Just as importantly, IE gives me the tools to advocate for better reproductive healthcare for Native women. This effort involves an arduous and often uphill battle that takes me to different reservation communities, the United Nations, and Washington DC, to work with nongovernmental indigenous organizations. I'd like to tell you these efforts result in immedi-ate improvements for Native women and their families. But change is hard, and it can take a long time. Mostly, I educate, and sometimes I argue. I share findings and offer assistance with disseminating these findings. I hope I empower, as I have been empowered by this research.

Most of my work, though, occurs on a local level. I frequently work with the resources that are immediately available to me—classrooms, students, campus organizations. For example, there are very few Native studies courses offered at the very large public university where I teach. This is where I come in; I lecture in numerous classes across campus every semester, using my sociological understanding of Indian Country to introduce liter-ally hundreds of undergraduate students to some of these complexities and some of the ways a sociological imagination can address them. At times, I am invited to speak at local high schools or for local organizations, where I share what Indian Country and sociology have taught me.

I've also worked closely with the Community Outreach Program on our campus to develop a service learning trip for undergraduates to the Pine Ridge Reservation, a trip which has since expanded to include a Cherokee reservation in Oklahoma, as well. A student who traveled with this pro-gram recently told me he's going back to Pine Ridge after graduation to work with young kids there, and another is seeking a job on the reserva-tion as a teacher. It makes me smile to think of the potential alliances being built, and new relatives getting to know each other, linked by sociological inquiry in this way. Many of the students with whom I work become active in the Native American Cultural Society (NACS) on campus, and some of them were instrumental in garnering (finally!) an office for the NACS. I recruit Native American speakers from across the county to conduct diver-sity workshops and I blog about issues relevant to Indian Country. I am able to do these things because my own sociological imagination and the sociological research which feeds it, have given me new insights and new relationships—and, with these, new responsibilities.

As an institutional ethnographer, I have learned a great deal about the social organization of all our lives, and particularly the ways in which our lives are frequently organized along the lines of race, class, gender, and citizenship. My ability to analyze the role of various institutions in producing and maintaining particular experiences helps me identify the strengths and weaknesses of American institutions, such as healthcare, and strategize about ways to improve these institutions. In these ways, sociology has helped me to be a better relative to Native people. I hope this brief essay about my work as a sociologist—work which impacts real lives—is of benefit for you. After all, we're relatives, as well.

References

Smith, D. E. (1987). *The everyday world as problematic.* Boston, MA: Northeastern University Press.

Smith, D. E. (2005). *Institutional ethnography: A sociology for people.* Toronto, ON: AltaMira Press.

PUTTING SOCIOLOGY TO WORK IN WINNERSVILLE, USA

Mark Patrick George

Valdosta State University, Valdosta, Georgia

Mark Patrick George is an applied sociologist from Valdosta, Georgia, where he currently serves as an assistant professor in the Department of Sociology at Valdosta State University. His scholarly attention and community organizing work focuses on antiracist, antisexist, and antiheterosexist initiatives in the Southeast. In addition to his academic pursuits, he serves as the faculty advisor for the Zeta Chapter of Alpha Kappa Delta and coordinates the Mary Turner Project (www.maryturner.org), a joint student/community organization created to addresses issues of racial injustice in South Georgia. He also serves as the education committee chairperson for the Lowndes/Valdosta Chapter of the Southern Christian Leadership Conference.

One of the things that attracted me to sociology was that it helped me make sense of the world I found myself in as a young white man growing up in the Deep South. So when I stumbled across sociology, as someone who had changed his major five times and didn't know why he was going to college, it was liberating. First, sociology showed me that it is good to ask critical questions and to question matters others accept as "normal." Second, sociology helped me find answers to questions that plagued me by teaching me to recognize the broader social forces that shape our lives—you know, that "sociological imagination" (Mills, 1959) professors are always going on about! For example, too often we simply fault poor people for being poor instead of understanding *why* they are poor, a practice William Ryan (1971) labeled "blaming the victim." Lastly, it freed me from an assortment of unnecessary fears and judgments I carried about others who are different from me by revealing that social differences are not inherently dangerous or "deviant" (Lorde, 1984), but in most cases are the spice that makes life interesting and vibrant. For these reasons, I became a sociologist.

While sociology was personally freeing, the courses I took rarely discussed what one was to do with this new understanding and way of seeing the world. Even less was said about how one might use sociological tools like research and theory to reduce injustice and unnecessary human suffering. So, for the past 20 years or so, those are the questions I have perpetually thought about and revisited. Today, I find myself thinking about those questions back where my sociological awakening began. That setting, a place of slow life and deep tradition, is Valdosta, Georgia. There I teach sociology courses, conduct community-based research, and work for social justice in the community.

Who's Losing in Winnersville?

Self-proclaimed by its citizens years ago to be "Winnersville, USA" because of its long tradition of high-school football championships, like many southern communities, Valdosta, Georgia, has an extensive yet rarely acknowledged history of racial terrorism and racial exclusion. Winnersville had to be forced by the U.S. government to desegregate its schools in the late 1960s and early 1970s, long after the *Brown v. Board of Education* Supreme Court decision of 1954, and was a place with a history of Ku Klux Klan activity and lynching. As one local activist put it, "Dr. King marched in Albany just 45 minutes from here, but the civil rights movement missed Valdosta altogether" (Dolores Brown, personal communication, July 25, 2003).

This legacy still haunts the community, and old-school forms of overt racism have simply mutated into more "civil," systematized forms of institutional and "colorblind" (Bonilla-Silva, 2010) racism today. In a county that is 62.7% white and 34.3% African American, and one where 32.2% of those African American families live below the federal poverty line, compared to 10.9% of white families (U.S. Census Bureau, 2009), whites continue to control most institutions and benefit disproportionately in Winnersville. In other words, not everyone is "winning" in Winnersville, and a variety of social indicators reflect this reality.

One of the most disturbing ways young African Americans are losing in Winnersville is educationally. As an alumnus of Valdosta City Schools (VCS), I have been monitoring local education data off and on for years, and today local schools are more racially segregated than they were 40 years ago. For example, while VCS's student population was 64% white and 46% African American in 1970, today only 18% of the VCS student population is white (VCS, 2008) and 77% is African American. After 40 years of "white flight" from Valdosta City Schools, this predominantly African American system reported a graduation rate of 50% for African American students in 2007–2008, while its minority white student population graduated at a rate of 78%, higher than the state average (Georgia Department of Education, 2008). In addition to an abysmal graduation rate among African American students, data I began to collect in 2007 reveal that individual schools within the VCS system are "internally segregated" in terms of where white and black students attend school, and that African American students are very underrepresented in advanced or accelerated academic programs[1] (i.e., gifted programs, honors programs, advanced placement programs, college preparatory courses). The data also reveal stark disparities between African American and white scores on all state and federally mandated standardized tests, and that African American students are grossly overrepresented in terms of retentions (i.e., students that are retained or fail their grade each year) and disciplinary actions[2] in VCS.

So, in a system that was federally forced to desegregate, African American students are falling through the cracks en masse in VCS. At the same time, there is a history of silence around this reality. Today, VCS remains a system controlled by white administrators and educators (a problem the U.S. Department of Justice has been monitoring and wants remedied) that works primarily for its white minority population.

---

[1]As of 2006–2007, black students made up only 40% of those in gifted programs, while white students, who are only 18% of the total student population, accounted for 54% of all gifted students in such programs (Valdosta City Schools, 2008).

[2]As of 2006–2007, African American students comprised 93% of out-of-school suspensions. In addition to suspending roughly a thousand African American students "out of school" annually (20% of the total VCS student body), J. L. Newbern Middle School (with a 98% African American student population) suspended 40% of its total student population (n=260) in 2008, and 80% of those suspensions did not involve violence or criminal activity but were simply defined as "other" by that school's administrators. To date, the Southern Christian Leadership Conference (SCLC) has been unable to secure accurate data on the number of students expelled each year in the VCS system.

Putting Sociology to Work and "Outing" Winnersville's Educational Crisis

The relationship between a person's education and life chances is a well-established fact. This relationship is also one that is visible in Georgia as a whole. As of 2008, 70% of Georgia's prison population (36,760 people) did not complete high school. Equally troubling is that, as of 2008, 63% (33,084 inmates) of that same prison population were African American (Georgia Department of Corrections, 2009) in a state where African Americans comprise only 30% of the statewide population (U.S. Census Bureau, 2009).

Having collected and compiled longitudinal data on the racial disparities in VCS, one of the ways I put social research "to work" in 2007 was to discuss these disparities with relevant community organizations. Since VCS administrators had been reluctant to share data with me,[3] interrogated me when I asked for information covered under Georgia Open Records Law, and repeatedly informed me that "race has nothing to do with any academic disparities," I turned my energy and attention to sharing my findings with those directly impacted by the data. I became involved with a local civil rights organization (the Lowndes/Valdosta Chapter of the Southern Christian Leadership Conference [SCLC][4]) and was asked to serve as Education Committee Chairperson. In that capacity, I began to share the data I had with the broader community through town hall meetings (Pinholster, 2008) and a Web site (www.winnersville.net). I was also asked to discuss the matter on local radio talk shows. In each of these venues, I was able to offer a sociological analysis of the data, discuss why this was a racial issue and not simply a matter of economic class, and to encourage others to employ that infamous "sociological imagination." Critical questions I and community leaders like Rev. Floyd Rose of the SCLC raised also included the following: Why had Valdosta City Schools become a predominately African American system? Had VCS always failed to serve students of color and, if not, what had changed? Why was VCS still under a 1970 U.S. Department of Justice order to fully desegregate? You get the picture. Because of these questions, much of the attention in the community shifted from simply blaming the victim to a more systematic analysis of VCS and the community at large.

---

[3]In 2008, I filed a formal complaint on behalf of the SCLC with the Georgia Attorney General's Office regarding VCS's noncompliance with the Georgia Open Records Act.

[4]The SCLC (Southern Christian Leadership Conference) was founded by Dr. Martin Luther King, Jr. and others in 1957 and was at the forefront of the civil rights movement.

My work with the SCLC around "who's losing in Winnersville" demonstrates the usefulness and power of social research and asking critical sociological questions. Although VCS remains racially problematic and under the Federal Consent Decree to address the lack of administrators and educators of color, noteworthy changes have begun to unfold. First, the resounding silence about educational racial disparities in Winnersville has been broken and has given way to ongoing public discussion and action. Second, the tendency for education administrators and the public to reduce VCS's failures to individuals has been superseded by a more systematic analysis of the system as a whole. That is evident in a number of new policies and programs developed by VCS in the fall of 2008 to address many of the academic disparities highlighted earlier. For example, as of 2009, J. L. Newbern Middle School and other schools are actively examining their out-of-school suspension rates and have publicly stated that they will work to cut those suspensions in half by 2010 (Georgia Department of Education, 2008). In addition, VCS as a whole has been more proactive about fostering parent and community involvement among its low-income student population by offering workshops and holding regular parent summits. Third, community members now regularly request relevant information from VCS, and VCS has begun to share more information with the community. Fourth, VCS School Board members have repeatedly and publicly been "put on notice" by community members to address the glaring academic disparities between students of color and white students in Valdosta schools. Fifth, VCS is more diligent about accurately documenting school board decisions and public input. Last, but certainly not least, over the last two years the public, particularly Valdosta's black community, has been involved and vocal in ways unlike anytime I have lived in Valdosta.

Putting Sociology to Work for Change

As I stated earlier, too often students of sociology are simply exposed to an assortment of social injustices without any ideas about how they might impact them. The work in Winnersville offers several possibilities, particularly in terms of how applied sociological research, in the right hands, can foster positive social change. I should also tell you that putting sociology to work will alter one's life for the better. Being a Sociologist in Action has meant that I have seen my work make a difference in the lives of others. That work has also enriched my life through the diversity of souls I have had the privilege of meeting, working with, and getting to know in the process. That's why I encourage all students of sociology to act on and apply what they learn instead of simply feeling helpless and powerless. I can assure you that your life will be richer and more vibrant for doing so.

References

Bonilla-Silva, E. (2010). *Racism without racists: Color-blind racism and the persistence of racial inequality in the United States.* Lanham, MD: Roman & Littlefield.

Georgia Department of Corrections. (2009). *Georgia Department of Corrections offender statistics.* Retrieved from http://www.dcor.state.ga.us/GDC/OffenderStatistics/jsp/OffStatsSelect.jsp

Georgia Department of Education. (2008). *J. L. Newbern School corrective action addendum.* Valdosta, Georgia: Author.

Lorde, A. (1984). *Sister outsider.* Berkeley, CA: Crossing Press.

Mills, C. W. (1959). *The sociological imagination.* New York, NY: Oxford University Press.

Pinholster, J. (2008, July 28). Town hall meeting raises questions about schools. *Valdosta Daily Times.* Retrieved from http://www.valdostadailytimes.com/archivesearch/local_story_210231452.html

Ryan, W. (1971). *Blaming the victim.* New York, NY: Pantheon.

U.S. Census Bureau. (2009). *State and county quick facts.* Retrieved from http://quickfacts.census.gov/qfd/states/13/1378800.html

Valdosta City Schools. (2008). *Unpublished data obtained from curriculum director and superintendent's office.* Valdosta, GA: Author.

METHODS OF TRUTH AND RECONCILIATION

David Cunningham

Brandeis University, Waltham, Massachusetts

David Cunningham teaches in the sociology department at Brandeis University. His research focuses on the causes and consequences of racial violence, with an emphasis on the civil rights–era Ku Klux Klan. At Brandeis, his community-based work has included a project on memory and oral history in the Mississippi Delta as well as the Possibilities for Change in American Communities program, which incorporates a month-long trip around the United States in a sleeper bus to teach and learn about historical and contemporary activist work in nearly two dozen communities.

The civil rights movement brought great change to Mississippi, which, like the rest of the South, had been using white supremacist Jim Crow customs and laws to maintain racial segregation into the 1960s. These segregationist practices not only separated white and black residents but also ensured that the latter had access only to substandard housing, education, and health care. African Americans were turned away from the most desirable jobs; were more likely to be harassed than protected by police; and were not allowed to eat, shop, or gather in public and commercial spaces favored by whites. This way of life was often brutally enforced. Even within the South, Mississippi was well-known for its racially oppressive practices. Private groups like the Citizens' Councils and the Ku Klux Klan intimidated anyone they deemed a threat to white supremacy. Government officials enabled and reinforced these actions, defying federal civil rights laws and establishing the Mississippi State Sovereignty Commission to promote segregation and investigate and prevent civil rights activity.

While much has changed since the 1960s, the legacy of Mississippi's racism remains evident. Neighborhoods and schools continue to be largely segregated—today, nearly half of the state's black students attend schools that are less than 10% white, and many white students continue to enroll in private academies originally established 40 years ago to circumvent school desegregation orders. Racial inequity is evident in other institutions as well. African Americans in Mississippi are more than four times as likely as their white counterparts to be incarcerated. The state's black citizens are chronically underserved by the health care system—more than twice as likely as whites to be uninsured and to report that they are in poor health. Salaries of black high-school graduates are, on average, 21% lower than whites with similar educational backgrounds. Black college graduates fair even worse, making 27% less than whites with a BA or BS degree.[5]

To come to grips with this connection between past discrimination and present-day racial inequality, a range of initiatives have focused on the vestiges of segregation. Recently reopened "cold cases" have led to the convictions of several former KKK members for past civil rights–related crimes.[6] Civil rights history is now a mandatory part of Mississippi's K–12

---

[5]For more on the segregationist academies established in Mississippi to avoid court school desegregation orders, see Andrews (2002). Data on racial disparities comes from Bureau of Justice Statistics (2001); Frankenberg, Lee, & Orfield (2003); Ruggles et al. (2004); Waidmann & Rajan (2000).

[6]A number of prominent Mississippi-based civil rights "cold cases" have been tried over the past 20 years. In 1994, the white supremacist Byron De La Beckwith was convicted of the 1963 murder of NAACP leader Medgar Evers. Ku Klux Klan (KKK) leader Sam Bowers was sentenced to life in 1998 for his role in the 1966 killing of voting rights activist Vernon Dahmer. In 2005, former KKK member Edgar Ray Killen was found guilty of manslaughter in the 1964 slayings of James Chaney, Michael Schwerner, and Andrew Goodman, who had been registering black voters in Mississippi as part of the 1964 Freedom Summer project. And in 2007, James Ford Seale was convicted on kidnapping and conspiracy charges for his role in a 1964 KKK-orchestrated plot to kill Charles Moore and Henry Dee in Mississippi's Franklin County.

school curricula, thanks to the 2006 Civil Rights Education Bill.[7] And an initiative called The Welcome Table promotes "an era of dialogue on race," by sponsoring a series of three-day retreats with community leaders to provide a foundation for participants to organize grassroots conversations and actions focused on racial justice.[8]

Perhaps the most ambitious of these efforts is the Mississippi Truth Project. Guided by a belief that "[a] just and inclusive future can only be ensured by a comprehensive inquiry of [the state's] unjust and segregated past," project participants are working to establish a statewide Truth and Reconciliation Commission (TRC) to "allow the state to constructively engage the confusion, division, and bitter feelings" that remain from earlier eras.[9] Following in the footsteps of postapartheid South Africa and more than two dozen other nations, the TRC will make use of public hearings, private interviews, and historical research to provide a space for both victims and perpetrators to tell their stories and for the community generally to work toward racial justice and reconciliation. Importantly, these efforts look forward as well as back. Connecting the dots between past discrimination and racial inequity in present-day Mississippi is how the Truth Project can, in the words of its organizers, "shape an inclusive and equitable future."

In 2009, my Applied Research Methods class at Brandeis University partnered with the Truth Project, enabling students to develop their research skills in support of social justice and social change. Located in New England, outside of Boston, Brandeis is a long way from Mississippi. But this work builds on the university's long association with civil rights efforts, which has included student and faculty involvement in the 1964 Mississippi Freedom Summer project and a South Carolina–based voter registration campaign the following year.[10] For this latest effort, our course took a research-team approach, with students, faculty, and community partners working collaboratively to meet grassroots needs. Our goal was to provide research support for the TRC by building a database and conducting analyses that allow its members to better understand and convey

---

[7]See Mississippi Senate Bill No. 2718, Regular Session 2006.

[8]The Welcome Table's structure and mission is described on its Web site, http://www .welcometable.net/.

[9]See http://www.mississippitruth.org.

[10]The 1964 Freedom Summer project brought more than 1,000 volunteers to Mississippi, mostly white college students from outside of the South, to assist with an effort to register black voters and establish a set of "freedom schools" for black youth (see McAdam, 1988). The following year, students from a smaller set of northern universities, including Brandeis, traveled to South Carolina to conduct voter registration work as part of the SCOPE project.

the institutional bases of support for the violence that emerged throughout this period. Members of the class worked directly with Susan Glisson, the executive director of University of Mississippi's William Winter Institute for Racial Reconciliation, which has provided financial and logistical support to the emerging Truth Project. Our efforts were collaborative, as Brandeis students had the opportunity to communicate with their counterparts in Mississippi, allowing us to better connect the seemingly abstract data we were collecting with more tangible real-world issues.

A key component of the Truth Project's research work is based in the recognition that a wide range of Mississippi institutions—including schools, banks, hospitals, businesses, police, and local governments—participated in the Jim Crow system. To support efforts to better understand the inner workings of segregation, each student focused on a particular Mississippi institution. Throughout the semester, students compiled an exhaustive synthesis of existing writings on their respective institutions, as well as gathered and analyzed original data focused on how these institutions operated differently in counties across the state. The class's collective project brought together these individual studies to construct detailed background reports as well as a master database integrating information on hundreds of factors tied to these institutions and their impacts on the African American population. In the process, topics that often fail to inspire passion in research methods courses—such as how we can confidently assume that research measures capture lived realities, how to select cases in a way that allows them to be treated as representative of broader populations, and so on—came alive in new ways as they informed the choices we made while constructing the TRC database.

At semester's end, the class sent its collective research product to Truth Project organizers. Dr. Glisson responded that she was heartened by the students' contributions:

> It was clear to us from the beginning that the Mississippi Truth Project would be served well by support and research from the academic community. . . . The issues being addressed are thorny and complicated and subject to attack from those who disagree; thus, having academic standing would strengthen the work. We have been amazed at the generosity and interest of those who see the value of grassroots leadership and who are working to provide resources to us.

Similarly, many students emphasized the satisfaction they gained from linking their research methods training to a real-world social justice initiative. "I have never felt 'proud' at the end of a course, and I feel that way after having taken this course," one student noted. While acknowledging the intensity of this sort of applied class, several participants stressed that their accountability to Mississippi-based community partners made the added effort worthwhile. "The class definitely required a substantial amount of work and commitment," a member of the class remarked, "but if you put in the time, the rewards are very much worth it." Another exclaimed, "The only suggestion I have is to make this course every single day, and one year long. I am very upset it is ending!"

But in a more general sense, Brandeis's involvement with the Truth Project hasn't ended. As the Mississippi TRC moves forward in the coming months and years, Brandeis students will continue to work collaboratively to apply their research skills to these issues of racial inequity. By applying the tools of sociological research to a complex problem, the class hopes to contribute to the TRC's ultimate goal: "to develop appropriate remedies and to create a culture of equity, harmony, and prosperity"[11] in Mississippi and beyond. As a sociologist, there is nothing more satisfying than linking students' passion for social justice to the tools of social science research, and witnessing the possibilities for change that result.

References

Andrews, K. T. (2002). Movement–countermovement dynamics and the emergence of new institutions: The case of "white flight" schools in Mississippi. *Social Forces, 80*(3), 911–936.

Bureau of Justice Statistics. (2001). *Prison and jail inmates at midyear 2001.* NCJ 191702. Retrieved from http://bjs.ojp.usdoj.gov/content/pub/pdf/pjim01.pdf

Frankenberg, E., Lee, C., & Orfield, G. (2003). *A multiracial society with segregated schools: Are we losing the dream?* Cambridge, MA: Harvard University Civil Rights Project.

McAdam, D. (1988). *Freedom summer.* New York, NY: Oxford University Press.

Mississippi Truth Project. (2009). *Declaration of intent.* Retrieved from http://www.mississippitruth.org/pages/declaration.htm

Ruggles, S., Sobek, M., Alexander, T., Fitch, C. A., Goeken, R., Hall, P. K., et al. (2004). *Integrated public use microdata series* (based on the 2000 Decennial Census): *Version 3.0* [Machine-readable database]. Minneapolis: Minnesota Population Center.

Waidmann, T. A., & Rajan, S. (2000). Race and ethnic disparities in health care access and utilization: An examination of state variation. *Medical Care Research and Review, 57*(1), 55–84.

---

[11]Mississippi Truth Project (2009).

DISCUSSION QUESTIONS

1. What prompted C.N. Le to become interested in his Asian American background and to eventually create Asian-Nation.org? If you were asked to create a sociologically based blog, what would be your focus? Why?

2. How does Le's blog and Web site help to both inform and support the Asian American community? What other purposes does it serve? Why do you think this site is unique?

3. How did you do on Barbara Gurr's Quiz for The Non-Native American? Why? What are some of the key reasons most non-Native Americans know so little about Native–U.S. relations?

4. One of the obligations of sociologists is to give voice to the marginalized in society. How does Barbara Gurr carry out that obligation? Is there a particular group in society for whom you would like to help "give a voice"? If so, which one? Why? If not, why not?

5. According to Mark Patrick George, who is winning in the public school system in "Winnersville"? Why? How is he using sociological tools to address the inequities in Winnersville?

6. How does the school system where you went to high school compare to that in Winnersville? What was the racial makeup of the schools, the advanced classes, and the lunchroom? Did you notice these racial trends when you were in school? If yes, why—and what did you think of them? If no, why do you think you did not notice them?

7. David Cunningham's students work to connect past racial injustices to present-day racial inequality. Do you think it is important to carry out such work? Why or why not? In what ways might their findings be helpful for those seeking to reduce racial inequality?

8. Why do you think Cunningham's students were willing to work harder than they might otherwise have to work in a regular research methods class? Would you be willing to put in extra effort in a course if you knew that your work would be used by those fighting for social justice? Why or why not? Is there a particular cause for which you would be willing to put in a great deal of extra work? If not, why not? If yes, which one?

9. How has reading the Sociologist in Action pieces in this chapter influenced your perspective on race relations in the United States? What information did you already know? What information surprised you? Why?

10. Describe how the work of at least two of the Sociologists in Action featured in this chapter exemplifies the two core commitments of sociology (use of the sociological eye and social activism).

RESOURCES

The following Web sites will help you to further explore the topics discussed in this chapter:

Ethnicity and Race Tutorial	http://anthro.palomar.edu/ethnicity/Default.htm
Internet Resources for Ethnic Studies	http://guides.lib.udel.edu/ethnicstudies
Leadership Conference	http://www.civilrights.org/index.html
Race: The Power of an Illusion	http://www.pbs.org/race/index.htm
Sociology of Race and Ethnicity	http://www.trinity.edu/~mkearl/race.html
Sociosite Ethnicity–Migration–Racism	http://sociosite.net/topics/ethnic.php
Understanding Race	http://www.understandingrace.org/

To find more resources on the topics covered in this chapter, please go to the Sociologists in Action Web site at **www.sagepub.com/korgensia2e.**

Chapter 10

Sex, Gender, and Sexuality

Sex, the physical attributes of males or females (or a combination, as in intersexed individuals); *gender,* the social roles assigned to males and females; and *sexuality,* how people express themselves sexually, are the foci of many Sociologists in Action. This chapter describes how some public sociologists have been addressing these issues through their research and how they conduct that research. The Sociologists in Action in this chapter have creatively used their sociological skills to work toward positive social change.

In "Sex in Some Cities: Explorations of AIDS/HIV Education and Hooking Up," Rebecca Plante colorfully describes how her experience interviewing a male, heterosexual dancer performing in a gay male club helped her to realize the complexity of sex and gender. In this piece, she also discusses her evaluative work on HIV/AIDS education and prevention efforts and her work with undergraduates conducting "interviews with college students . . . to gain some clarity about what 'hooking up' means." Plante has helped her students realize that examining sex and gender issues from a sociological perspective helps them understand both themselves and their society more thoroughly.

Michael Kimmel succinctly shares his findings from his book *Guyland* (2008) in "A Public Sociology of Gender and Masculinity." Through conducting more than 400 interviews of young men between the ages of 16 and 26, Kimmel was able to discover "that many forces conspire together to keep [young men] from developing a life plan, a mental map of where they want to go and how to get there." Kimmel used the term *Guyland* for this new stage of development before adulthood. By illustrating its "sociological, or structural, foundations," he helps "young people . . . navigate its treacherous waters more consciously and more ethically." Kimmel has consulted with leaders of fraternities, governments, high schools, and institutions of higher education to help make the communities within these institutions more positive for everyone and to encourage men to work for gender equality.

In "The Southern West Virginia PhotoVoice Project: Community Action Through Sociological Research," Shannon Bell lucidly illustrates how her use of feminist participatory action research has enabled her to understand injustices in the coalfields of southern West Virginia and to help the subjects of her research (40 women from coal mining communities in the region) to become active in confronting them. Bell's PhotoVoice Project used "participant-produced photography and narratives, coupled with group reflection, to facilitate 'empowerment education' among participants and provide a venue for voicing their concerns." Giving the participants control over the subject matter and providing a forum for their photostories to be seen and discussed by people with political power helped them effectively voice their community's concerns and hopes for positive change.

In the final piece in this chapter, "Getting the Message Out," Susan Stall vividly describes how she and a colleague (Roberta Feldman) embarked on a 10-year action research project with women residents of Wentworth Gardens, a public housing development in Chicago, Illinois, working to protect their homes from demolition for the new Chicago White Sox Stadium. The academic highlight of their research was the publication of *The Dignity of Resistance: Women Residents' Activism in Chicago Public Housing* (Feldman & Stall, 2004), and their work also helped bring about a positive result for the residents of Wentworth Gardens. Stall believes that their findings and efforts to publicize the work and struggles of the group of women fighting for their housing played an important role in the successful effort to save Wentworth Gardens.

References

Feldman, R. M., & Stall, S. (2004). *The dignity of resistance: Women residents' activism in Chicago public housing.* New York, NY: Cambridge University Press.

Kimmel, M. (2008). *Guyland: The perilous world where boys become men.* New York, NY: HarperCollins.

SEX IN SOME CITIES: EXPLORATIONS OF AIDS/HIV EDUCATION AND HOOKING UP

Rebecca Plante

Ithaca College, Ithaca, New York

Rebecca Plante is an associate professor of sociology at Ithaca College and has taught at Wittenberg University, University of New Hampshire, and Tufts University. Her classes focus on gender and sexuality, beauty, culture, and theory. Dr. Plante is the author of *Sexualities in Context: A Social Perspective* (2006), and she is the coeditor of *Sexualities: Identities, Behaviors, and Society* (2004, with Michael Kimmel) and *Doing Gender Diversity* (2009, with Lis Maurer). She has also worked as a (peer) sexuality educator and HIV/AIDS program evaluator. From 1997 to 1998, "Latex & Vinyl," a radio call-in show at the University of New Hampshire, featured Professor Plante (as "Dr. Victoria Monk") answering listeners' questions about sex.

In a small bar in Lexington, Kentucky, I watched as a muscular guy gyrated and danced in a tiny, lime green bikini. He stood on a small stage, covered with baby oil, shining under the spotlight, and thrusting his pelvis at the appreciative crowd. He was part of a dance troupe billed as the "Nine Inch Males," and as part of my postdoctoral research (study undertaken after a doctoral degree is completed), I was there to administer very brief surveys to patrons. When the (somewhat) erotic and eventually fully nude dance was done, he made his way to where I stood, pencils and questionnaires ready.

"What are you doing here?" he asked.

Up close, I could see the faintest hint of the stubble on his arms and chest; his green bikini was more like a thong—nothing covered his behind.

"I'm doing research," I replied, glad to have an obvious answer, as I gestured to the study materials in front of one of the bartenders.

As we talked, I learned a few things. First, he was as young as he looked—23, it turned out. He had no idea what "postdoctoral research" was but correctly assumed that I was affiliated with the nearby flagship state university. It turned out that he was working his way through college and that he considered himself to be heterosexual. Given that he had just performed for a happy and affectionate group of men crowded into the small gay bar where we stood, I admit that I was a little surprised to hear this.

"I used to dance for women—bachelorette parties, 21st birthdays, sorority events—but honestly, the men tip way better. It's not a big deal because I just think about the money." Our conversation forced me to do some thinking that influenced my research immediately, and led to a major change in my subsequent understanding of sexuality and gender-related issues.

My postdoctoral research involved a multisite, mixed-method evaluation and assessment of HIV/AIDS education and prevention efforts. I collected survey, interview, and focus group data about Kentucky's attempts to reach everyone from gay men to injecting drug users to men and women in underserved communities (namely, people who are often left out of the health care system—immigrants, working-class people, etc.). In doing so, I called on my sociological tools and training. I used my sociological training to write appropriate survey questions for questionnaires sent to the county health departments, trying to capture their perspectives on the services they provided and the people they served. My sociological knowledge about the functions of bureaucracies and social institutions came in handy as I read and analyzed the returned surveys, and would later prove invaluable in meetings with stakeholders, the groups of people with investments (tangible or not) in the outcomes of my evaluation research. And my awareness of how individuals create selves within social contexts was indispensable as I conducted interviews and led focus groups in men's prisons, with gay men living in rural areas, and with HIV-positive people.

After crisscrossing the state and administering hundreds of surveys, facilitating 10 focus groups, and conducting another 30 interviews, I was ready to present the findings to the stakeholders, including professionals at the state and county health departments, people living with AIDS and HIV infection, and my research supervisor—in short, several rather diverse and complex groups. Successful communication required me to draw on my sociological imagination, which has given me an appreciation for the complexity of social concerns, along with an ability to see how personal issues and public troubles intersect. For example, through focus groups I learned that some physicians in Kentucky were routinely but wrongly telling HIV-positive patients that they could never be sexually active again. If this had happened to just one person, it could have been considered a "personal trouble," a problem between someone and his doctor. But it happened to multiple people living in multiple towns, and since sexuality is central to many people's lives, the kind of information that doctors give their patients about sex can become a "public issue," especially if it is incorrect.

Though HIV/AIDS education and evaluation work is important, and it enabled me to influence how such education was provided (at least in Georgia and Kentucky, where I have done evaluation studies), I was curious about other kinds of sexual issues in people's lives. Sex is indeed central to many people's lives, and with that in mind, I have tried to think as a sociologist might when studying how we develop a "sexual self." Imagine trying to apply a big-picture lens to something that seems as intimate, individual, and personal as sexuality. It is not an easy task but is a sociologically important one.

Together with a team of undergraduate researchers, I have been attempting to understand more about how we become sexual—how we learn, adapt, and grow—given the broader social and historical contexts in which our sexualities exist. Recently, we've been studying *hooking up*, a vague term for a bunch of activities ranging from kissing to making out to oral sex to intercourse. The team has been doing interviews with college students, trying to gain some clarity about what hooking up means. More importantly, we are trying to learn how people feel about hooking up, whether they think hooking up is the only way to be sexually involved with someone, and what it really means to say that it's a fun thing to do. We're also trying to find out if hooking up differs based on certain variables, such as gender, sex, race, ethnicity, and sexual orientation. For example, what explains why some people hook up less than others? Do gay and lesbian students surrounded by mostly heterosexual students hook up less, as they may have fewer people to do it with?

It is too early to know if our study will have any public policy implications, or if it will ever be read by powerful stakeholders (as in the case of my AIDS/HIV evaluation research). But we already know that our findings can be applied in a variety of ways. One simple but often-overlooked "application" is that the study has led to people talking about hooking up in ways that transcend Twitter updating, texting their friends, and talking over dinner. When I read our interview transcripts, I see the trained undergraduate interviewers bringing things to the surface that invariably result in interviewees saying, "I never thought about that before." In a meaningful way, our interviewees are thinking critically about their sexualities and why they are who they are. Some talked about hooking up almost as a habit, like teeth brushing, without much thought about the whys and hows. At the end of many interviews, people commented on whether hooking up was sexually or emotionally satisfying (and for some, it was neither).

One member of the research team reflected on her experiences:

Researching hooking up gave me an intimate look at people's inner world, and allowed them to talk about private matters that they have always wanted to discuss but may not have felt able to. It allowed me to look at my own life critically but not judgmentally. Sex is private and vulnerable, but in our culture it has become so taboo or fetishized that it almost feels ethereal, unreal. Studying sex allowed those who talked about it to really connect to their stories, and helped me see connections between all people, on an intimate level and in a respectful way.

The study has also sparked the creation of several different college courses—a small-group tutorial on intimacy, an upper-level class on the sociology of intimacy, and a one-week intensive course for advanced high-school students on hooking up. The students in these classes brought our readings, discussions, and commentary to further and more diverse audiences—they talked to their friends, the people they hook up with, and their boyfriends and girlfriends. One student sent me an e-mail at the end of the semester:

> Before I enrolled in college as a freshman, I considered myself an empowered, sexually liberated woman. Logically, one could ask why. I believed that my ability to "hook up" with whoever I wanted and under my terms (with a complete disregard for the other person's feelings) meant that I was an individual—which as a woman must mean that I was empowered. As my years in college went by, I realized that I was sabotaging my friendships and romantic relationships by regarding myself as "just" an individual. The biggest consequence was that when I was hurt after a hook-up I believed I was the only one who was hurting and that everyone else was otherwise "empowered." A pivotal moment for me in the Hooking Up class was listening to other women who also voiced similar feelings of loneliness, being used, and aloneness.

Throughout the Hooking Up class, students talked—about themselves, their experiences, their neighborhoods, their families, their friends, their peers, their older siblings, their media. But they went beyond straightforward recitations of their own opinions and actions—they applied theories, methods, and a sociological imagination to hooking up, love, intimacy, and sexualities. My students came to see that studying hooking up—something seemingly so individual and unique—is a way to be more innately sociological. One student used her summer internship to design a pamphlet like those in campus student health centers. She wanted to encourage people to talk about their feelings about hooking up and to become comfortable with their choices and decisions.

By being more sexually, romantically, and intimately aware, we become people who have ethics and who seek justice, in our own bedrooms and for other people. As one student wrote in an end-of-semester self-evaluation,

> I came to realize that we are socialized in this country to believe that we should be ashamed of our sexuality and that we are individuals. The worst part about

this is that we are presented with a dark paradox when we realize that our access to sexual rights (i.e., abortion, sterilization, rape, consent) as women is constantly being evaluated as a whole. I could have stopped here, and given up and said, "Well, everyone feels shitty and it's just part of life." But that would mean that I would silence the voices of all the women who I represent when I speak up about the troubles we all face in the bedroom, on dates, in marriage, etc. If you asked me, ["]What did you take away from the Hooking Up class?["] I'd say,["]I had the privilege to hear and see what it means to exist within an empowered sexual community.["]

And what about the "Nine Inch Male" in the lime green thong? What was the major change in my own understanding of sexualities that followed after our meeting? I finally understood that the best and most useful tool for a sociologist of sexuality to possess is a constant appreciation of the beautiful complexity of sex and gender. The work I do is sometimes incremental and occasionally maddeningly individual, but I know that we all benefit from talking about sexuality, intimacy, bonds, and connections—especially in a thoughtful, in-depth way.

A PUBLIC SOCIOLOGY OF GENDER AND MASCULINITY

Michael Kimmel

State University of New York (SUNY) at Stony Brook

Michael S. Kimmel is a professor of sociology at SUNY at Stony Brook. He is the author of *Guyland: The Perilous World Where Boys Become Men* (2008), which was featured on the *Today Show; Good Morning America;* and over 100 radio, newspaper, and blog reviews. His other books include *The Politics of Manhood* (1996), *Men's Lives* (8th edition, 2009) *Manhood: A Cultural History* (1996), and *The Gendered Society* (3rd edition, 2008). He also co-edited *The Encyclopedia on Men and Masculinities* (2004) and *Handbook of Studies on Men and Masculinities* (2004), and is the founder and editor of *Men and Masculinities,* the field's premier scholarly journal. He lectures extensively in corporations and on campuses in the United States and abroad.

For several years, colleagues have been describing a growing gap between male and female students on their campuses. The women, they observed, seemed so focused, so organized. They had their whole life planned out—how they'd graduate, get a good job, begin to search for a mate and start a family. They not only could envision their lives 10 years after graduation, but they could tell you how old they'd be when they have their first child.

As for the men . . . well, not so much. They seemed adrift, confused and anxious about their futures, unable to envision their lives in 10 minutes, let alone 10 years. Underneath the affable sociability of "it's all good" lay some roiling uncertainty.

What a paradox! On the one hand, any observer could tell you that men are still "in power," that it's still a man's world. Look at any corporate board, national or local legislature, or municipal government. Women's wages still lag behind men's, and there's still plenty of discrimination, harassment, and violence that shapes women's lives. Sexual assault seems epidemic. But on the other hand, here are college women who seem to be running circles around the men, outnumbering and outperforming them on campus, gathering honors the way I used to collect baseball cards.

This was the paradox I set out to explore in my book *Guyland* (2008). As a sociologist who studies gender—and particularly masculinity—I knew the answers were not simply rooted in different psychological dispositions of women and men. It wasn't just immaturity or extended adolescence on the part of males. I suspected it might be more that many forces conspired together to keep them from developing a life plan, a mental map of where they wanted to go and how to get there. Through the more than 400 interviews I conducted of young men between the ages of 16 and 26, I was able to discover that there is now a new stage of development between adolescence and adulthood, a stage of development that has its own shape and dynamics. I call it *Guyland*, and it has sociological, or structural, foundations:

• The demographic revolution—Young people in the developed world will live longer than any generation in world history. Why marry and commit to a career at age 22, like their parents, only to be married for 75 years? (Women cannot usually delay childbearing into their 40s, so they cannot delay adulthood as readily.)

• Economic changes—The increased instability of economic life, global economic competition, increased domestic competition, and the abdication of corporate responsibility toward employees and employees' loyalty to the company, have made committing to a career more tenuous and terrifying than ever before.

• Parenting changes—The changes in parenting, which I term *helicopter parenting*, in which parents micromanage every nanosecond of their children's lives and intervene for their "special" children at every moment, thus creating children—boys and girls—who please adults and lack resilience.

- Changes for women—The dramatic transformation in women's lives, as women have entered every field formerly closed to them, become far more sexually adventurous, and become professionally ambitious.

It's in this context that young men struggle to prove themselves as men. Sometimes they look back nostalgically at an earlier time, when men were men and women hadn't invaded those formerly all-male bastions where men could become "a man's man," a "man among men." (Think of the hit TV show *Mad Men*.) Some drift, seemingly aimlessly, in a world now anomic and alienating. And some go to extreme lengths to prove themselves, taking excessive risks, drinking way too much, desperate to prove themselves. Theirs is not a Peter Pan syndrome, but more like a Peter *Panic* syndrome.

As a public sociologist, my job is to make these findings available to a larger public, rather than simply to other researchers and scholars. For several years, I was a contributing editor at *Psychology Today* where I wrote a column called "Women and Men," covering gender issues. These days, I lecture all over the world, write for newspapers and magazines, and have a regular blog on the *Huffington Post.*

Politically, I see my work as making a contribution to women's struggle for gender equality. Specifically, my job is to show empirically that supporting women's equality will actually benefit men. Just as men are not from Mars and women are not from Venus, so too is gender equality *not* a zero-sum game in which if women win, men will lose. The evidence points decidedly in the other direction: Here, on planet Earth, women's equality will enable men to live richer, fuller, happier lives—longer and healthier lives animated by better relationships with their friends, partners, spouses, and children.

The task I set for myself in writing *Guyland* was to chart this world, understand both the origins of this new stage of development between adolescence and adulthood—a stage that affects both women and men—and to map Guyland as a social space, the arena in which this new stage takes place, an arena characterized by indeterminacy, anxiety, and often some defensive resistance. I believe that while this new stage of development is here to stay—young people are extremely unlikely to return to earlier models of achieving adulthood by their early 20s—young people can navigate its treacherous waters more consciously and more ethically. There is a role for parents, for friends, and for the community at large to ease the transition, and to interrupt the constant pressure young guys feel to prove themselves to be "real men." If "it takes a village to raise a child," it takes a nation to guide those children toward responsible adulthood.

Through writing *Guyland,* I have researched the world my students live in, taught about that world, and conducted activist work to encourage others to engage with young men at this critical period of their lives. I've consulted with governments, colleges and universities, and high schools about making those worlds safer and more supportive for everyone. Some national fraternities have embraced my critique of Guyland and are devising plans to eliminate hazing. I've also developed programs to engage men as more active fathers, to challenge bullying in school, and to involve men in the struggle for gender equality.

It's been through sociology—the field that invites us to contextualize our experience, that requires that we connect biography to history, that encompasses both the problem and the possible solution—that I have found my calling to be engaged. We're often told that as scholars we have to "leave our politics at the door," to be as dispassionate and disengaged as possible. I say, "Nonsense." I think that unless we are committed and engaged with the world, our work will be lifeless and our lives emptier. But, at the same time, unless we commit ourselves equally to intellectual rigor, we will be pundits preaching to an increasingly shrill choir. A public sociologist needs to have both something to say, and the analytic and methodological tools to be able to say it convincingly.

Reference

Kimmel, M. (2008). *Guyland: The perilous world where boys become men.* New York, NY: HarperCollins.

THE SOUTHERN WEST VIRGINIA PHOTOVOICE PROJECT: COMMUNITY ACTION THROUGH SOCIOLOGICAL RESEARCH

Shannon Elizabeth Bell

University of Kentucky, Lexington

Shannon Elizabeth Bell is an assistant professor in the sociology department at the University of Kentucky. Her research falls at the intersections of gender, environmental sociology, and social movements, and she maintains a strong commitment to feminist research and social justice in her work. Bell received a PhD in sociology and a graduate certificate in women's and gender studies from the University of Oregon in 2010. She incorporated PhotoVoice into her dissertation to examine the barriers to local participation in the Central Appalachian environmental justice movement, which is fighting to hold the coal industry accountable for irresponsible mining practices.

As one of the top coal-producing regions in the United States, Central Appalachia's political, economic, and social landscape has historically been—and continues to be—closely tied to the coal industry. Dubbed an "internal colony" of the United States by numerous scholars (Lewis, Johnson, & Askins, 1978), Central Appalachia has long been exploited for its resources in order to fuel the nation's energy demands.

Over the past 20 years, this exploitation has reached new levels as an extremely destructive form of coal mining, known as *mountaintop removal,* has become widespread throughout Central Appalachia. In this form of coal extraction, companies blow apart mountains in order to expose and remove a thin seam of coal. This form of mining, coupled with other industry practices, has led to a great deal of flooding in nearby communities (Flood Advisory Technical Taskforce, 2002), respiratory disorders from coal dust (Ohio Valley Environmental Coalition, 2003), coal waste impoundment collapses (Erikson, 1976), and groundwater contamination (Orem, 2006). For many years, residents of the Appalachian coalfields have been unfairly forced to bear these extreme environmental injustices in the name of cheap energy for the rest of the nation.

The increasing frequency of coal mining–related flooding, sickness, and water contamination in Central Appalachia has led to the emergence of a grassroots-based environmental justice movement in this region. Coalfield activists are demanding protection from and accountability for the destruction and pollution in their communities through confronting the coal industry; regulatory agencies; and local, state, and national governments. While the movement was started in large part by local residents (especially women) fighting to protect their families from the dangers associated with coal mining and processing, most of the affected population remains uninvolved in the environmental justice movement. Especially in recent years, activists have been forced to recruit social movement participants from outside the region, expanding into noncoal mining areas and even non-Appalachian states.

Given the enormity of the impacts of irresponsible mining practices on these communities, why is there such limited movement participation at the local level? What are the barriers to grassroots mobilization in Central Appalachia? And what are possible mechanisms for overcoming those barriers?

My field research spanned 13 months in the coalfields region of southern West Virginia, where I conducted in-depth interviews, participant observation, content analysis, surveys, and a PhotoVoice project with 40 coalfield women. Most often used in human service fields, PhotoVoice uses participant-produced photography and narratives, coupled with group reflection, to facilitate "empowerment education" among participants and provide a venue for voicing their concerns (Wang & Burris, 1994). As a feminist participatory action research (PAR) method, PhotoVoice provides data for the researcher while simultaneously benefiting the research participants.

In September of 2008, I initiated an eight-month PhotoVoice project with 40 women across five different coal mining communities in southern West Virginia. I provided participants with digital cameras and asked them to take pictures to "tell the story" of their communities, including the problems and the positive aspects of life. I facilitated monthly meetings in each of the five communities over the course of the eight months, providing an opportunity for the women to share their photographs, discuss common themes, create photostories (photographs with written narratives), and openly discuss and question the underlying structural causes of the community problems they identified. As a part of the project, the women held public exhibits and presentations of their photostories, communicated their communities' concerns and major issues to state legislators, and developed projects to initiate change in their communities. As a researcher, I studied the process whereby many of the women began to see themselves as agents of change in their communities and participants in the democratic process.

Feminist critiques of fieldwork have pointed to the power imbalances that exist in both the design of research and the research setting. Participatory action research projects such as PhotoVoice attempt to challenge these power dynamics by viewing research participants not as subjects but as partners in the study and cocreators of the design, implementation, and benefits of the research. Thus, in my own project, while I was interested in studying the social processes taking place during the course of the PhotoVoice project, the *participants* had power over the ultimate outcomes of the project itself. They decided what to photograph, what aspects of life to write about, and which photostories they wanted to include in the public exhibits and presentations; they became their own documentarians with the power to decide how their lives and communities should be represented, instead of allowing outside photojournalists to make those decisions for them. By providing a venue for their photostories to be publicly viewed and discussed with people in positions of political power, the PhotoVoice process also allowed participants to convey their communities' concerns and their ideas for positive change to a wider audience. Thus, the benefits of the research were mutually shared among the researcher and the research participants.

This project had different purposes for the various women involved. Many focused their efforts on dispelling negative stereotypes about their region, showcasing the breathtaking beauty of southern West Virginia's mountains, creeks, and wildlife. Others chose to document important cultural traditions and the rich history of the area through their photostories. Some used their photostories as a way to directly communicate their concerns and ideas to legislators and others with political power.

Those who chose to use PhotoVoice as a tool for political involvement focused on a number of different issues. Many of the women revealed the devastation the coal industry is causing in their communities through creating photostories documenting mountains being blown apart, scarred landscapes, and toxic water. Others only minimally focused on the coal industry, instead choosing to concentrate on other issues, such as dilapidated and dangerous roads, or the overwhelming amount of trash littering the roadsides and creek banks.

As certain themes emerged within the five PhotoVoice groups, I encouraged the women to write, call, and schedule meetings with their state legislators. I also provided the women with a list of bills being considered by the West Virginia legislature that addressed some of their largest concerns. One of these bills was a bottle deposit law. For the seventh year in a row, the West Virginia Legislature was considering a bill that would place a three-cent deposit on all beverage containers sold in the state. Many women across the five groups mailed their legislators photostories depicting the large number of plastic bottles, glass bottles, and aluminum cans littering their communities. In their narratives, they told why they believed a bottle deposit would help with this problem. Many of the women were surprised by the positive reactions they received from their legislators. For instance, when one of the participants from the Harts PhotoVoice Group received a return phone call from her state representative asking if she could tell him more about why she wanted a bottle deposit in the state, she was shocked. While the bottle deposit bill did not pass during the 2009 legislative session, the experience of being *heard* had an empowering effect on the participant (as well as many of the other women). Reflecting on her efforts with the bottle deposit bill through the PhotoVoice project, this participant revealed,

It hasn't happened yet, but it made me think that it was possible. You know? A lot of times whenever [I] get a big dream or idea in [my] head, I've never really thought real positive about it. It gave me a positive outlook on it anyway, you know, instead of being such a downer.

A more immediate success story took place in the Pond Fork area of Boone County, where a number of the PhotoVoice women created photostories revealing the crumbling and dangerous roads in their community. In one of our regular monthly PhotoVoice meetings, two of the participants expressed that they would like to go to the State Capitol and talk to some of their legislators about the road conditions in their area. I gave them contact information for their state senators and house members, and they called the capitol and scheduled appointments with two of their representatives.

We put together packets of information for their visit, including full-color copies of all of the photostories the group had created that related to the road conditions, as well as a CD of homemade video footage taken while driving down the pothole-covered roads.

The day of their appointments, I met the PhotoVoice women at the capitol, and we went to their legislators' offices. The photostories, video footage, and these individuals' heartfelt descriptions made quite an impression on the two legislators. Both legislators made promises that they would see what they could do to have the roads in the Pond Fork area moved up on the list of repaving projects. In the meantime, both made phone calls to the Department of Highways during our visit to get the roads patched as a temporary fix. Later that same day, the Department of Highways was in the Pond Fork area patching the roads that the PhotoVoice group had photographed! Since then, one of the women has been in regular contact with her legislators, checking on the progress being made toward full repaving of their roads.

The photostories that the PhotoVoice women created are continuing to have an impact. We have printed booklets of photostories depicting coal-related water pollution problems in the coalfields. These booklets are being used by various citizen organizations in the state to educate legislators about why it is important to pass a moratorium on underground coal-waste (slurry) injection operations.

More than 400 of the photostories created by the women involved in the Southern West Virginia PhotoVoice Project can be viewed at our online gallery at www.WVPhotoVoice.org.

References

Erikson, K. T. (1976). *Everything in its path: Destruction of community in the Buffalo Creek flood.* New York, NY: Simon & Schuster.

Flood Advisory Technical Taskforce. (2002). *Runoff analyses of Seng, Scrabble, and Sycamore Creeks, Part I.* Division of Mining and Reclamation, Department of Environmental Protection. Retrieved from http://www.wvdep.org/Docs/1593_Part%20I.pdf

Lewis, H. M., Johnson, L., & Askins, D. (Eds.). (1978). *Colonialism in modern America: The Appalachian case.* Boone, NC: Appalachian Consortium Press.

Ohio Valley Environmental Coalition. (2003). *Coalfield residents speak.* Retrieved from http://www.ohvec.org/issues/mountaintop_removal/articles/2003_12_07_EIS_speakanon.pdf

Orem, W. H. (2006, November 15). *Coal slurry: Geochemistry and impacts on human health and environmental quality.* U.S. Geological Survey, Eastern Energy Resources Team. PowerPoint Presentation to the Coal Slurry Legislative Subcommittee of the Senate Judiciary Committee, West Virginia Legislature.

Wang, C., & Burris, M. A. (1994). Empowerment through photo novella: Portraits of participation. *Health Education Quarterly, 21*(2), 171–186.

GETTING THE MESSAGE OUT

Susan Stall

Northeastern Illinois University, Chicago

Susan Stall was a union representative for the United Teachers of Los Angeles, while pursuing her MA in sociology of education, working on the issues of metropolitan desegregation and bilingual education. Since then, she has worked with numerous community groups including Women United for a Better Chicago. Dr. Stall's Presidential Address to the Midwest Sociological Society in 2009 was entitled, "Civic Sociology." She was a founder and co-chair of her university's Applied Learning and Engaged Scholarship Committee with the goal of significantly expanding faculty engagement in service-learning and community-based action research across the campus.

[I]t was decided and approved that we would choose the name—Wentworth Gardens Residents United for Survival—because we felt at that point that's what we really would be. Because here were the White Sox right up on us threatening to take our home, and here's CHA [Chicago Housing Authority] doing nothing for us in here. It was the only name that really fit what we were about to be about!

—Mrs. Hallie Amey, president of Wentworth
Gardens Residents United for Survival

It was the threat to the physical survival of Wentworth Gardens, a low-rise public housing development, and its surrounding neighborhood brought on by the proposed new Chicago White Sox stadium that brought Wentworth Gardens residents into the political arena to engage in battle to protect their homes. A Chicago, Illinois, newspaper article, in late 1986, showing the plans for the new Comiskey Park, which required the demolition of part of the Wentworth Gardens development, was their impetus to take action. According to Mrs. Amey, reading that article and map is "what made everybody's hair stand up on end!" That's when women activists, at Wentworth Gardens, one of 19 family public housing developments in Chicago, decided to fight the proposed new baseball stadium. It was shortly after this point that I first learned of these activists, their battle, and their newly formed nonprofit organization—Wentworth Gardens Residents United for Survival.

In 1986, while teaching as a part-time instructor at the University of Illinois at Chicago, I was hired to be an organizer for a unique nonprofit, Women United for a Better Chicago (WUBC). I secured funding from WUBC to organize two-year-long grassroots efforts to bring together diverse groups of individuals and organizations that were working on the issue of providing safe and affordable shelter and housing for women and their families (Stall, 1986). These organizing efforts culminated in two 2-day conferences: Women and Safe Shelter: Creating and Recreating Community, in April of 1986, and Women and Public Housing: Hidden Strength, Unclaimed Power, in April of 1987.[1] As I conducted research for the latter conference, I became acquainted with some of the key activists at Wentworth Gardens, and I asked them to make a presentation about their efforts in the conference session, "Planning Our Communities to Resist Displacement." I was impressed with their organizing acumen, their positive energy, and their history of successful on-site achievements to create and secure their community.

I continued my part-time organizing with WUBC and worked with a committee of public housing residents from across the 19 family CHA developments, and with resident advocates, to form the citywide tenants' organization, Chicago Housing Authority Residents Taking Action (CHARTA) (Kagan, 1991). Through this effort, I came to even better appreciate the Wentworth activists. I also partnered with another academic practitioner, Roberta Feldman, an environmental psychologist and architectural educator. Together, we visited the development several times and gathered some individual and group interviews with the intention of writing a short article about the residents' challenges around the White Sox battle, and to briefly document some of their past achievements for a grassroots magazine, *The Neighborhood Works*. We felt it was important to get the word out to foundations and the organizing community about the courageous and impressive efforts of these residents (Feldman & Stall, 1989).

---

[1]At the Women and Public Housing Conference, local and national community activists living in or working with public housing explored such issues as leadership development, self-employment, resident management, and welfare reform in 25 conference sessions. Wentworth Gardens appears in the document that accompanied the conference, *A Resource Directory for Residents in Public Housing*, which identified 130 on-site service providers and advocates for public housing residents and included 12 case studies of resident activists (Hunt-Rhymes, Oppenheim, Stall, & Young, 1987).

As Roberta and I came to know these resident-activists better, we were struck by their creativity, camaraderie, and tenacity in the face of formidable challenges. We became increasingly convinced that it was important to offer an alternative portrayal of public housing residents—one that contested existing pervasive negative stereotypes. Oft-cited studies of life in public housing depict residents as on the defensive, attempting to protect themselves from surrounding and increasingly internal human and physical threats (Newman, 1972; Popkin, Gwiasda, Olson, Rosenbaum, & Buron, 2000; Rainwater, 1970). More generally, social scientists typically characterize low-income people as helpless and apathetic victims of despair (Kieffer, 1984; Naples, 1988; see also Rappaport, 1981).

Yet I found myself questioning if two white, professional, middle-class women were the appropriate spokespersons for these African American low-income activists. But as Roberta pointed out, who else was going to tell their story and perhaps attract the needed professional and material resources they needed to support their organizing efforts? Thus, we entered into the adventure of a 10-year-long sociology Action Research project.

Our research topics were suggested and inspired by a core group of Wentworth Gardens activists and the advocates who worked with them. The activists' interviews and organizing efforts helped us to uncover their key historical achievements, their current noteworthy initiatives, and their acknowledged community partners and adversaries. The residents defined our research agenda, and at several points they also helped gather data and served as effective spokespersons in a wide assortment of venues. We were cognizant of our responsibility not only to reciprocate the Wentworth residents' contributions to our research but also to contribute to positive social changes in their community whenever possible.

There were several ways in which my role as a sociologist and my social capital as a professional were of particular value to the residents. Shortly after we began our research at Wentworth, in 1989, I was invited to organize and chair the Wentworth Gardens Resident Management Community Advisory Board. The advisory board was made up of professionals including community organizers, a housing lawyer, an architect, urban planners, social workers, and a few other local professionals. The tasks of this advisory board were to advise and provide technical and organizing support for Wentworth's new resident management board.

Throughout the 1990s, we attempted to promote the efforts of resident activists by documenting and giving public recognition to their past and current organizing achievements in a variety of alternative venues. As an example, from 1994 through 2000, Wentworth activists were featured in a one-hour segment, "Low-Income Women's Resistance," of a 24-hour telecourse called *Women and Social Action*. The segment takes the viewer to

Wentworth Gardens in a 12-minute clip where key resident activists describe organizing successes, such as their on-site laundromat and grocery store, and challenges, such as their deteriorating grounds. In addition, I am interviewed to explain the theoretical importance of the residents' community-building actions. *Women and Social Action* was distributed nationwide (live feeds as well as videotapes) by PBS Adult Learning Services from 1994 until 2000, which meant that it could be viewed on local PBS stations nationwide.[2]

In 1994, Roberta and I also prepared an exhibit for the Chicago Peace Museum that provided photo documentation and text (from our research) describing community-based actions at Wentworth. Moreover, Wentworth Gardens residents participated in several conference and colloquium presentations with me, and at times with other academics or organizers, as we discussed their activism through slide documentation and discussions in varied settings and for diverse audiences.[3]

The academic culmination of our research was the publication of our book, *The Dignity of Resistance: Women Residents' Activism in Chicago Public Housing* (Feldman & Stall, 2004). The resident activists featured in the book were excited about its publication, but two other public events were of more personal importance to them. The first was in 2005, when Wentworth resident-activists traveled to San Francisco, California, to receive the National Social Advocacy Award from the American Planning Association, an award that "recognizes the plans, practices, people, and places that further the field of planning and help create communities of lasting value" (http://www.planning.org/awards/2005/index.htm).[4] The

---

[2]The accompanying study guide to the course, *Women and Social Action: Teleclass Study Guide* (Thompson, 3rd edition, 2008), describes and analyzes the activism of the Wentworth Gardens residents, as does one of my earliest articles with Roberta Feldman, "Resident Activism in Public Housing, a Chicago Public Housing Development: A Case Study of Women's Invisible Work of Building Community" (Feldman & Stall, 1990).

[3]Since 1988, one or more Wentworth Gardens residents have presented with me in each of five colloquium or conference settings.

[4]The description in the American Planning Association Program read, "National Social Advocacy Award, *Wentworth Gardens, Chicago, Illinois.* Struggling against the odds, a group of women activists at Chicago's Wentworth Gardens have undertaken a 40-year campaign to keep their community on the radar screen of public officials. The challenges at the 1,200-resident housing complex are manifold: building code violations, millions of dollars in deferred maintenance, and under-funded recreational and education activities. To call attention to these concerns, leaders of the Wentworth Gardens' Resident Management Corporation and local Advisory Council in the 1960s began a sustained grassroots effort to remedy the most pressing problems. Among their victories were: pressuring the Chicago Housing Authority to commit $1 million for renovating the complex's dilapidated field house, reopening a community field house, and stopping an attempt to demolish the housing complex in order to make way for a new White Sox stadium."

residents were featured at the National Planning Association's Annual Awards Luncheon, where Mrs. Amey, one of the resident-activists, offered some brief words of acceptance and appreciation to a room full of thousands of professional planners and academics. Even more exciting for the residents, and the advocates who had worked with them for decades, was the play adaptation of our book, *A Neighborhood Fight: A Play in Two Acts*. Through this cultural medium, residents had the exciting opportunity to observe actors dramatically portray them and their organizing efforts on stage. To date, there have been two widely attended public performances of this play. The first was the featured entertainment at the 10th anniversary celebration of the Neighborhood Writing Alliance on April 24, 2006.[5] The second play performance was at the renowned Chicago Humanities Festival, an annual series of lectures, concerts, films, and cultural events, on November 4, 2006; the theme of that year's festival was appropriate for the Wentworth struggles: "Peace and War: Facing Human Conflict."[6]

The fact that Wentworth Gardens was not destroyed, as the majority of the public housing development units in Chicago have been in the last 10 years, is due to the residents' continual vigilance, creative community-building efforts, and effective social bridging to supportive advocates. Yet, I also believe that our longitudinal research, and particularly our efforts to help publicize the residents' "dignity of resistance"—their formidable barriers and achievements—have helped to attract the government redevelopment monies and foundation organizing dollars needed to secure their safe and affordable family housing. Throughout my work as a Sociologist in Action with the Wentworth resident-activists, I felt privileged to be welcomed into such an emotionally warm and vibrant community. The lessons I learned from these women about the dignity of resistance and the importance of "women-centered organizing" (see Stall & Stoecker, 1998) have deeply informed my teaching, research agenda, and my own activist efforts ever since.

---

[5]There was an accompanying story about the residents in the fall issue of *American Planning Association* magazine, 2005.

[6]In 2004, I approached a local award-winning playwright, David Barr III. He expressed interest, and for several months I met regularly with David and his cowriter, *Chicago Tribune* journalist Glenn Jeffers, to provide background information beyond the book. When the play came out, I, along with the residents, had the strange experience of seeing "Susan Stall," an activist sociologist, played by an actress on stage!

References

Feldman, R. M., & Stall, S. (1989, June/July). Women in public housing: "There just comes a point . . ." *The Neighborhood Works,* 4–6.

Feldman, R. M., & Stall, S. (1990). Resident activism in public housing: A case study of women's invisible work of building community. In R. I. Selby, K. H. Anthony, J. Choi, & B. Orland (Eds.), *Coming of age: The proceedings of the 21st International Environmental Design Research Association Conference* (pp. 111–119). Urbana-Champaign, IL: International Association of People–Environment Studies.

Feldman, R. M., & Stall, S. (2004). *The dignity of resistance: Women residents' activism in Chicago public housing.* New York, NY: Cambridge University Press.

Hunt-Rhymes, K., Oppenheim, L., Stall, S., & Young, G. (1987). *Women & public housing—hidden strength, unclaimed power: A resource directory for residents in public housing.* Chicago: Women's Studies Program, University of Illinois.

Kagan, S. (1991, June 2). Public housing tenants organizing for self-help. *New York Times.* Retrieved from http://www.nytimes.com/1991/06/02/us/public-housing-tenants-organizing-for-self-help.html

Kieffer, C. H. (1984). Citizen empowerment: A developmental perspective. In J. Rappaport, C. Swift, & R. Hess (Eds.), *Studies in empowerment: Steps toward understanding action* (pp. 9–36). New York, NY: Haworth.

Naples, N. A. (1988). *Women against poverty: Community workers in anti-poverty programs, 1964–1984.* Doctoral dissertation, City University of New York (University Microfilms International, Ann Arbor, MI).

Newman, O. (1972). *Defensible space.* New York, NY: MacMillan.

Popkin, S. J., Gwiasda, V. E., Olson, L. M., Rosenbaum, D. P., & Buron, L. (2000). *The hidden war: Crime and the tragedy of public housing in Chicago.* New Brunswick, NJ: Rutgers University Press.

Rainwater, L. (1970). *Behind the ghetto walls: Black family life in a federal slum.* Chicago, IL: Aldine de Gruyter.

Rappaport, J. (1981). In praise of paradox: A social policy of empowerment over prevention. *American Journal of Community Psychology, 9*(1), 1–26.

Stall, S. (1986). Women organizing to create safe shelter. *Neighborhood Works, 9*(8), 10–13; *9*(9), 10–12.

Stall, S., & Stoecker, R. (1998). Community organizing or organizing community? Gender and the crafts of empowerment. *Gender & Society, 12*(6), 729–756.

Thompson, M. E. (2008). *Women and social action: Teleclass study guide* (3rd ed.). Dubuque, IA: Kendall Hunt.

DISCUSSION QUESTIONS

1. Rebecca Plante helps her students examine "hooking up" from a sociological, rather than an individualistic, perspective. In the process, they learn more about (a) society, (b) what influences how they behave sexually, and (c) how they think about sex. How do you think looking at hooking up from a *sociological* perspective helps them in these ways? How did reading this piece help you to think sociologically about sex and sexuality?

2. How did the man in the lime green bikini influence Rebecca Plante's understanding of sexuality and gender issues? Can you imagine yourself carrying out such an interview? How do you think your own gender socialization might influence your ability/willingness to conduct such an interview?

3. According to Michael Kimmel, what are the sociological foundations of *Guyland*? What are some signs of the existence of this new stage of development on your own campus? How do you think your college can take steps to address many of the negative issues now associated with this new stage of development? How can students help to accomplish this endeavor?

4. If you are a young adult male, do any of Kimmel's observations hit home? If so, which ones? If not, why not? How can a study such as Kimmel's guide how parents raise their children? How has it influenced the way you view males in the 16–26 age range? Why?

5. How does Shannon Bell's research project rely on feminist methods of collecting data? What are the advantages to (a) the participants and (b) the researcher of using such a methodology? How might you use this type of data collection to research a social problem and to help create social change?

6. Were you surprised by any of the motivations of the women who participated in the PhotoVoice project? If so, which ones? Why? If not, why not? If you were asked to participate in such a project about your hometown, what would be your purpose in participating? What would you hope to gain from it? Why?

7. Susan Stall and her colleague worked with the women of Wentworth Gardens for over 10 years. What do you think made this long-term project (a) possible and (b) worth undertaking for both the residents and the researchers?

8. How were residents of Wentworth Gardens involved in shaping Stall and her colleague's research project? If you were a researcher carrying out a study, would you feel comfortable having your respondents participate in the research process? Why or why not? What do you think would be gained from this, and what would be some of the downsides?

9. How can a sociological perspective help us recognize that norms of sexual behavior are socially constructed (and therefore change over time and from society to society)?

10. Describe how at least two of the Sociologist in Action pieces in this chapter relate to each other in terms of either their subject matter or their research methodologies.

RESOURCES

The following Web sites will help you to further explore the topics discussed in this chapter:

ASA Sex and Gender Section	http://www2.asanet.org/sectionsexgend/
ASA Sociology of Sexualities	http://www2.asanet.org/sectionsex/
GLBTQ Encyclopedia	http://www.glbtq.com/
History of Women's Suffrage	http://www.history.com/topics/the-fight-for-womens-suffrage
National Organization for Women	http://now.org/
Sociosite Feminism and Women's Issues	http://sociosite.net/topics/women.php
Sociosite AIDS and HIV	http://www.sociosite.net/topics/health.php#AIDS
Sociosite Sex-Gender and Queer Studies	http://sociosite.net/topics/gender.php

To find more resources on the topics covered in this chapter, please go to the Sociologists in Action Web site at **www.sagepub.com/korgensia2e.**

Chapter 11

Globalization and Immigration

Globalization and immigration are two "hot button" issues often in the headlines these days. *Globalization,* the increased connection and interdependency of societies throughout the globe, and *immigration,* the movement of people from one society to another, are closely tied to one another. This chapter contains stories of Sociologists in Action who strive to understand and positively influence the impact of these issues in the United States and across the globe. Their efforts help us to use our sociological eye to look beyond the headlines and notice social patterns that affect the fabric of our everyday lives and our society.

Joe Bandy, with Elspeth Benard (Dennison), starts off the chapter with "Interracial Conflict and Attempts at Reconciliation in Auburn, Maine." This piece describes how Bandy provides opportunities for his students to use the sociological skills they are learning to help organizations in the community address a social problem. Bandy and Benard focus on Benard's honors thesis through which she (1) examined the causes of interracial conflict among the immigrant and native youth of Auburn, Maine, and (2) contributed to efforts toward interracial reconciliation by evaluating a "Controversial Dialogues" program at an area high school. Benard's findings allowed her to make several recommendations that were implemented to promote racial tolerance and improve race relations in the community.

In "How Culture Matters in Poverty Alleviation Efforts: Microcredit and Confucian Ideas in Rural China," Becky Hsu describes how she used sociological tools to show the impact of culture on poverty alleviation work. Through her research, she was able to explain why two different microcredit programs have resulted in very dissimilar repayment rates, and show "that financial incentives are mediated by different kinds of social environments." These findings reveal that those who wish to alleviate poverty must be aware of the impact of culture on behavior and adjust their programs according to the norms of local cultures.

In the third piece in this chapter, Jackie Smith describes how she has used her sociological tools to help challenge the corporate dominated globalization process. Her participatory observation research on global activism has taken her from protests against the World Trade Organization (WTO) in the streets of Seattle to World Social Forums around the globe. Through this participatory research, she has helped activists from various backgrounds form collaborative relationships, fostered efforts to strengthen diversity and democracy within the global activist movement, and "promoted a different story about globalization—one that stresses the negative consequences of global markets for both working and unemployed families around the world." In the process, she has helped spread the word about, and inspired many others to join, the global justice movement.

As do most public sociologists, Irene Bloemraad likes to extend her teaching and research skills beyond the traditional classroom. In "Informed Debate in a Political Minefield," she shares stories of helping organizations assist immigrants more effectively and of dealing with community members who are riled up about immigration issues. Whether she is carrying out a community presentation or conducting a workshop for lawyers and immigration service providers, she pushes her audience to examine immigration issues from a social scientific perspective and think through evidence about the various issues with which they are dealing. In the process, she helps members of the larger community to become more informed and educated citizens.

In "Community-Based Research and Immigrant Rights," Leah Schmalzbauer describes her work on "a community-based research project to document, analyze, and address the most urgent needs of the Montana Mexican community." Schmalzbauer's collaborative research efforts have resulted in a wide variety of positive repercussions for the Latino community in Montana and have expanded our collective knowledge of immigration issues. She concludes that, while there is still a long way to go toward achieving "security, opportunity, and justice for all community members . . . the collaborative research, teaching, and organizing efforts of which [she has] been a part are making a difference." Her work is a shining example of a sociologist having an impact on an important social issue.

INTERRACIAL CONFLICT AND ATTEMPTS AT RECONCILIATION IN AUBURN, MAINE

Joe Bandy, with Elspeth Benard (Dennison)

Vanderbilt University, Nashville, Tennessee

Joe Bandy was an associate professor of sociology at Bowdoin College from 1998 until 2010 and is currently the assistant director of Vanderbilt University's Center for Teaching. His research and teaching have focused on topics that include economic development in Latin America, environmental justice and labor movements, transnational movement coalitions, as well as poverty and class inequality in the United States. He and his students have partnered with many organizations in Maine to research local social and environmental problems related to poverty and development.

Elspeth Benard (Dennison) received her BA in sociology from Bowdoin College in 2006. Since graduating, she has worked for the Boston Private Industry Council (PIC) where she runs academic and professional development programs for underserved Boston teens. Part time, she is earning a master's of divinity from Gordon-Conwell Theological Seminary's Center for Urban Ministerial Education, which she hopes to employ serving internationally.

In my teaching at Bowdoin College, I encourage my students to be publicly engaged around important social issues by using a variety of teaching methods, including participation in original research projects that are community-based. I provide my students the opportunity to conduct original research designed in a collaborative process by me, the students, and a community nongovernmental organization chosen carefully by the students. Typically, the projects attempt to study some dimension of a social problem or policy, yielding information that will assist the organization to better understand and respond to the community's needs. In this manner, each student has the opportunity to become a "sociologist in action," enjoying the chance to actually *do* sociology by applying theories and methods to practical and publicly relevant issues while assisting a community partner.

One example of such a community-based project was an honors thesis that was conducted by Elspeth ("Ellie") Benard (now Dennison), a Bowdoin College senior in 2006, under my supervision. Ellie had a deep commitment to understanding and resolving interracial and interethnic conflict. She was particularly interested in how these conflicts are being shaped by effects of globalization, such as migration, economic restructuring, terrorism, and war.

The case of Lewiston and Auburn, two neighboring towns in central Maine, presented an ideal opportunity for social research and community development. Lewiston-Auburn is a 96% white community of approximately 65,000 residents claiming predominantly French, English, Canadian, and Irish ancestry. Since 2001, the community has experienced an influx of over 1,500 Somali immigrants, most from the Bantu minority ethnic group. The new Somali residents arrived as secondary migrants, refugees from the Somali civil war in the 1990s who originally resettled in larger American cities, such as Chicago, Illinois, and Atlanta, Georgia, but who subsequently moved to smaller towns seeking cheaper and safer communities.

Both whites and non-whites in Maine have suffered from economic hardship in recent years, most notably job insecurities related to deindustrialization. The result of the economic downturn in Lewiston-Auburn, coupled with the arrival of the Somali immigrants, created a classic scenario of interracial competition that, in combination with anti-immigrant fears post-9/11, was likely to ignite into full-scale resentment given the right spark. This spark came in the form of an infamous open letter to the Somali community in 2002 by Lewiston's former mayor, Laurier T. Raymond.[1] In the letter, which was published in the local paper, he urged Somalis to stop moving to the area because they had "overwhelmed" the city's services and "maxed out" its finances. These baseless but incendiary claims ignited a firestorm of public outcries and protests, including an anti-immigrant rally supported by out-of-state white supremacists, and counterdemonstrations for racial unity and tolerance supported by many of Maine's grassroots groups and political leaders.

Although overt tensions have subsided in recent years due to the painstaking work of many local and state organizations, racial and ethnic resentments continue to spill into occasional hostilities, including the troubling recurrence of racial violence in area high schools between whites and Somalis, and between whites and African Americans. In one such school, Edward Little High School in Auburn, school administrators needed assistance as they tried to encourage intercultural understanding and acceptance. They turned to the Center for Preventing Hate (CPH) (formerly the Center for Hate Prevention). CPH is a nonprofit founded in 1999 with the mission of "developing and implementing training, educational and advocacy programs to prevent bias, harassment and violence."[2]

CPH had recently started an initiative, "Controversial Dialogues," designed to facilitate interracial dialogue among the most contentious

---

[1]A copy of the letter can be found at www.americanpatrol.com/INVASION/ME/Ltr2SomalisME_Mayor021005.html
[2]See the CPH Web site at www.preventinghate.org.

students in area schools. After some discussion with school administrators, CPH introduced this program into Edward Little High School in 2004. While its first year of implementation demonstrated noticeable successes, CPH wanted to improve their program based upon an independent evaluation.

This provided an opportunity for my student, Ellie, to develop an honors research project that helped her to bridge sociological research with the strong possibility of influencing real-world outcomes. She was able to design a research project that enabled her to outline the complex causes of interracial conflict among Auburn's youth *and* make a contribution to interracial reconciliation by assessing the effectiveness of CPH's Controversial Dialogues program. Working with the assistance of CPH, Ellie gathered information through three different methodological techniques: participant observation, focus groups with students, and surveys administered before and after the program. Throughout the 2005–2006 school year, Ellie sat in on several Controversial Dialogues sessions among students and took notes about what she saw and heard, particularly the ways students appeared to be understanding and processing the origins of racial difference and conflict. In small-group discussions, or "focus groups," she asked students to evaluate the program and its educational value around issues of race. Through the surveys she had students take before and after the program, she compared their levels of racial consciousness before and after their participation in the Controversial Dialogues program.

Ellie's data indicated that broader forces of interracial and interethnic competition in central Maine, forces that were largely external to the school itself, were causing the school's white youth to develop prejudicial understandings of the differences between blacks and whites and Mainers and Somalis. Particularly evident were the fears of economic vulnerability among many of the white students' parents, coupled with a tendency to scapegoat racialized immigrant groups like the Somalis, as unfair competitors for jobs and as burdens on public funds (as Mayor Raymond had done). Ellie's findings also showed that these prejudiced beliefs were leading to the open hostility and sometimes violent conflict between students. Her data also indicated that white parents feared interracial dating among their children and Somalis.

At the same time, the data revealed that Somali experiences with overt racism were creating defensive reactions among both Somali parents and children in Auburn. Here, fears of prejudice and conflict caused Somalis to insulate themselves from their white peers, resulting in estrangement and misunderstandings. These economic and familial dynamics, on both sides of the racial divide, were creating a culture of racial animosity and fear that, at times, spilled over into open confrontation and violence, including arguments and fights.

As CPH had recognized, one solution to this racial conflict rested in having the most prominent and antagonistic members of the racial and ethnic groups engage in carefully structured dialogues to better understand their fears, histories, and cultural differences. These dialogues were designed to diffuse the most extreme racial and ethnic fears while promoting intercultural understandings. The anticipated result would be for participants to move toward developing antiracist identities forged through a more critical awareness of the historical and social structures of racial privilege. From these antiracist identities, workshop participants would develop a stronger commitment to a more integrated and multicultural society.

Ellie's evaluation of CPH's Controversial Dialogues program revealed that it was successful in helping students to consciously understand the racialized nature of their conflicts and the ways racism negatively impacts the lives of whites and non-whites alike. This certainly helped to lessen the contentious climate at Edward Little and reduced harassment and bullying, while providing students with an open forum to air grievances and diminish racial biases. Further, some students were more likely to commit to interracial conflict resolution after the CPH program, suggesting some growth of antiracist identities among the participants. On the other hand, Ellie's study revealed that students, particularly whites, did not develop a more mature understanding of race as a structure of privilege, nor of the ways in which their language or actions were implicated in this structure. Therefore, their understanding of racism remained limited. They did not commit to unmasking the many covert ways that racial privilege and inequality are maintained. Moreover, students who participated did so inconsistently because of scheduling conflicts with other school responsibilities. Lastly, Ellie uncovered that there were no opportunities for faculty or administrators to participate in a similar program or to develop curriculum and schoolwide events around lessons on racial conflict.

Consequently, Ellie made several recommendations to both CPH and Edward Little High School. These included rescheduling the dialogues so as not to conflict with classes or extracurricular obligations, establishing faculty/ administrative antibias education, and integrating critical race studies into the standard curriculum in ways that help students engage maturely with racial conflict in all of its personal and public manifestations. Ellie also unearthed, through her extensive literature review and secondary research, examples of other programs around the nation that have implemented one or more facets of these recommendations. This provided CPH and Edward Little High School with useful models, including the pioneering antiracist work of Beverly Tatum, author of *"Why Are All the Black Kids Sitting Together in the Cafeteria?"* (first published in 1997). At last report, the CPH's staff and board were integrating many of these recommendations into their programs.

Ellie's service and research experience with the CPH was crucial in shaping her own racial and social consciousness, and continues to affect her daily work. She is now a program coordinator for the Boston Private Industry

Council (PIC). At the PIC, Ellie runs academic enrichment and employment programs for inner-city, minority youth, for whom systemic forms of racism are an all too persistent reality. About this work, Ellie has said,

> It is designed to empower underprivileged youth with education and then channel them into stable careers in prominent Boston industries such as health care and financial services. It helps to break down racialized barriers to achievement and makes the city's socioeconomic landscape more racially just. Each summer, aspirations for meaningful employment in famous gleaming buildings and for a hard-earned dollar are realized by 4,000 Boston students through PIC programs. (Personal communication, April 9, 2009)

While this educational and job program does not represent the end of racism, to Ellie and others at PIC, it helps to transform racial concepts of "self" and "other" that too often stunt minority aspirations and lead to racial inequality. For Ellie, thorough research into the causes of racial inequality and effectiveness of antiracist programs, whether in Maine high schools or in Boston's large labor market, is one important part of the foundation for greater racial justice. For me, Ellie's experience represents the best of publicly engaged sociology, since it resulted in problem-based research that empowered her to inform and improve community efforts for greater racial tolerance and equality. Indeed, her research is a great example of a Sociologist in Action, having used the tools of her sociological training to truly make a difference.

HOW CULTURE MATTERS IN POVERTY ALLEVIATION EFFORTS: MICROCREDIT AND CONFUCIAN IDEAS IN RURAL CHINA

Becky Hsu

Georgetown University, Washington, DC

Becky Hsu is assistant professor of sociology at Georgetown University. Her research interests include China, religion, economic sociology, organizations, and health. Her current research projects include one on the definition of happiness in China and another on the role of religious congregations and social capital in reducing health disparities in Washington, DC. She has examined efforts to solve poverty in rural China, estimating Muslims worldwide, international religion data methodology, and faith-based organizations. She received a BA in sociology with *cum laude* and distinction in the major from Yale University, MA with distinction and PhD from the sociology department at Princeton University.

I have used sociological methods to understand poverty alleviation efforts in the context of globalization, as development programs from certain regions of the world are "exported" to others. *Microcredit,* for instance, is the practice of giving very small loans to poor people to help them increase their incomes (some people use their loans to start businesses). Because many poor people cannot provide material collateral, microcredit often uses group lending and "social collateral," a method of incentivizing loan repayments with the use of social pressure. The enormous interest in microcredit has been fueled by the prominence of the Grameen Bank in Bangladesh and its founder, Muhammad Yunus (2003), the joint recipients of the Nobel Peace Prize in 2006. Currently, microcredit is being attempted in places as diverse as Sri Lanka, Nepal, Mexico, Bolivia, Peru, Egypt, Kenya, Uganda, Mongolia, Cambodia, and the United States.

In my work, I have used the sociological perspective to answer the question of how cultural differences influence the way people respond to poverty alleviation efforts. When I learned about microcredit and its use of social pressure to motivate loan repayment, I wondered if people around the world apply and respond to social pressure in the same way, or if cultural determinants might lead to different responses. Research suggests that there are different cultural and social patterns around the world. For example, specific behaviors are appalling in one culture while they are fine in others: cooking in clothes worn also to defecate is considered wrong by Brahmans, while Americans do not have this cultural belief (Shweder & Bourne 1984). Similarly, experiments show that people process information differently: shown the same underwater scene, Japanese people noticed aspects of the background like rocks in the bottom, while Americans noticed elements in the foreground like the largest fish (Nisbett, 2003).

I decided to do my dissertation research in a county in rural China. Confucian ideas are important in traditional Chinese thought, and in Confucianism, person-to-person obligations are more important than person-to-group obligations. In the county, there were two existing microcredit programs, each one using different social collateral structures. I was curious to see how the borrowers would respond to the two different programs. I used methodological *triangulation*—the use of several sociological methods—in hopes of working toward a more thorough understanding of the dynamics and effectiveness of each form of social collateral.

The first method I used was *ethnography* (a method using observation and interviews in field sites with the purpose of describing a group of people), through which I observed, firsthand, how the programs were administered. I spoke with staffers and followed them as they did their work, and I uncovered something important: The two programs were structured differently in terms of how the social pressure was supposed to work. The first made repaying the debt a person-to-group obligation, in that all the borrowers in the village had to repay their loans in order for anyone else to get a subsequent loan. Therefore, people were expected to face pressure from the entire village to repay. In contrast, in the second a guarantor was elected in each village—if anyone failed to repay, the guarantor would do so on the person's behalf. Then, that borrower would owe the guarantor (a person-to-person obligation). In all other capacities, the programs were very similar, thus making it easier for me to focus in on the effectiveness of these two different avenues to social collateral.

The second method I used was *quantitative*. I gathered records from the microcredit programs and looked for patterns in repayment. From doing this, I found that less than two-thirds of the loans were repaid in the first (person-to-group) program, while everyone paid their loans in the second (person-to-person) program. Loan repayment is important because it ensures that the programs can keep running, and therefore is often used as a measure of the success of microcredit programs.

Finally, I also conducted *interviews* with 100 villagers to gather personal experiential data about social patterns in daily life and to ascertain their response to the microcredit programs they were in. I found a common pattern of emphasizing person-to-person (rather than person-to-group) obligations in daily life. This corresponded with what they said about the two programs. Overall, in my interviews I learned that respondents were hesitant about, and not as compelled by, the person-to-group obligation requirements in the first program; this would help to explain why far fewer loans were repaid. In contrast, respondents from the second program reported being quite comfortable with the way the program was set up, based on person-to-person obligations, and thus always repaid their loans.

In cultural analysis, the goal is to compare different types of observable phenomena to see if there are any common patterns. I aimed to compare what I found through the three different research methods to explain what was going on in the field site. I found that there were two programs with dramatically different repayment rates, and that the response to the two programs is consistent with the way that villagers structure their social relationships in daily life, as well as their philosophical ideas about social relations in traditional Confucian thought. The findings showed that financial incentives are mediated by different kinds of social environments. Although the borrowers in this field site need the money just as badly as people living in poverty in other areas of the world, they conduct their social relationships by different rules. These differences change the way that poverty alleviation efforts must be practiced in different areas of the world. Therefore, as globalization allows people from different regions to come into contact with one another, we need to be cognizant of how culture matters, even when talking about financial incentives. People may have good intentions in funding and trying to administer microcredit programs around the world, but they must be done with an understanding of local culture and social patterns. Sociological methods are critical to having an accurate understanding of local social patterns.

Through conducting this research, I have had the opportunity to spend a lot of time with people born into completely different circumstances from mine. Seeing very different social patterns has forced me to question assumptions and see that culture has a tremendous effect on social life. Most of all, it was satisfying to know that my research efforts could help people design better poverty alleviation programs around the world.

References

Nisbett, R. (2003). *The geography of thought: How Asians and Westerners think differently . . . and why.* New York, NY: Free Press.

Shweder, R. A., & Bourne, E. J. (1984). Does the concept of the person vary cross-culturally? In R. A. Schweder & R. A. LeVine (Eds.), *Culture theory: Essays on mind, self, and emotion* (pp. 158–199). New York, NY: Cambridge University Press.

Yunus, M. (2003). *Banker to the poor: Micro-lending and the battle against world poverty.* New York, NY: Perseus.

LOCALIZING INTERNATIONAL HUMAN RIGHTS: ENGAGING WITH THE WORLD SOCIAL FORUM PROCESS

Jackie Smith

University of Pittsburgh

Jackie Smith is professor of sociology at the University of Pittsburgh. She is active in Occupy Pittsburgh and in the United States Social Forum's National Planning Committee. She has authored, coauthored, or co-edited several books on transnational social movements, including *Social Movements in the World-System: The Politics of Crisis and Transformation* (2012, with Dawn Wiest), *Social Movements for Global Democracy* (2007), *Global Democracy and the World Social Forums* (2007, co-authored with 11 others), and *Coalitions Across Borders* (2004, co-edited with Joe Bandy).

During my high-school years in the 1980s, the United States government was escalating Cold War rhetoric and the nuclear arms race while engaging in deadly military interventions in Central America. In response, a wave of peace activism spread across the United States, and adults in my church and in my school helped introduce me to international peace and justice activism. I helped send supplies to Nicaragua and organize resistance to the nuclear arms race, and when I graduated high school I applied to universities with programs in peace studies. I wanted to use the tools of social science to understand the causes of violence and the sources of peace and social justice. In particular, I wanted to learn why the peace movement in the United States had little impact on U.S. foreign policy, despite having widespread popular support for many of its goals.

Since the time I began my research, and as the world has become increasingly globalized, many more social movements have become active in international contexts, and many more groups are organizing transnationally (across nations) to reduce economic inequality and advance human rights and environmental justice. In the late 1990s, I began using *participatory observation research* as a method to study the movements I wanted to better understand. I found that it was impossible to fully grasp the strategic thinking and the complex and dynamic aspects of transnational organizing without becoming more deeply involved in the work itself.

A participatory research approach has benefitted my research in numerous ways, and I often learn a great deal more from the activists and movements I study than I can from scholarly works alone. As they develop strategies of action, movement participants articulate their own theories of social change. By doing participatory research, we can bring these sources of knowledge into our scholarship while also bringing scholarly analyses, skills, and energy to the work of social movements.

Understanding how power operates requires an exploration of what voices are not expressed and what options are not explored. Why is it that certain groups tend to have their preferences realized while others do not? What is it about the visions or preferences of weaker groups that is incompatible with the interests of the powerful? Engaging our sociological imaginations when answering these questions can help us identify alternatives to existing practices that contribute to various social problems.

Participatory Research on Global Justice Activism

By late 1999, it was becoming clear to many activist groups that the World Trade Organization (WTO) and the larger processes of global economic integration it represented were huge threats to the environment, human rights, and democratic institutions. Countries joined the WTO and other international trade agreements to improve their position in the global economy, and by doing so they relinquished considerable authority in the realm of economic policy. Members of the WTO agree to enact policies that facilitate international trade and investment, even if these policies contradict national laws and preferences. WTO rules tend to favor the interests of transnational corporations and the most powerful states because their large economies and economic resources allow them disproportionate influence (see Smith and Moran 1999). This has meant that national regulations that privilege products for their health benefits that are produced in ecologically friendly ways or by local businesses have been deemed "illegal" under WTO rules. For example, the WTO rules undermine international codes against misleading baby food advertisements, overrule restrictions on trade in seafood that is caught in manners threatening to dolphins, turtles, and other sea life, and prevent local communities from passing laws banning trade with countries under authoritarian rule (Wallach and Woodall 2004).

Many of the tens of thousands of activists who gathered in Seattle for the "Battle in Seattle," against the WTO in 1999 were critical of its undemocratic features and of its social and environmental consequences (see Smith 2001). They called for a different kind of globalization that puts human rights and environmental sustainability above profit-making, and they helped launch what has become a growing global justice or "alternative globalization" movement. A key complaint of people in this movement is class inequality linked to economic integration. They point out that, according to the United Nations (UN) Development Program, in 1960 the world's wealthiest 20% earned just 30 times the incomes of the poorest 20%. By the late 1990s, incomes of the richest 20% approached *80 times* that of the poorest 20%. This inequality is both a cause and an effect of the extraordinary role corporations and other financial players play in national and global politics. Because governments have been encouraged to focus on economic growth, they have tended to focus their support on the needs of businesses. Thus, while specific claims of people in this movement vary, most participants demand greater democracy and economic equality.

One of the first challenges the global justice movement faced was to convince the public that economic globalization as framed by the WTO and economic elites was neither inevitable nor natural. They needed to show that alternatives to industrial mass production and globalized trade were viable and able to meet people's needs. The mass protests in Seattle drew attention to how trade policies advanced a particular set of class interests, opening the way for people to question many basic assumptions about globalization. Many observers were surprised to see so many people in a country of such privilege resisting economic globalization. Thus, while it was not the first mass protest against a global financial institution, Seattle became a turning point in the history of global justice activism. Following the Battle in Seattle, large numbers of activists began gathering more regularly to protest international meetings on global trade and finance.

I attended the protests in Seattle and subsequent demonstrations against the World Bank and Free Trade Area of the Americas in order to learn how activists from different countries and sectors (such as labor, students, Indigenous, anti-poverty, feminist) were able to find a common sense of purpose and coordinate their actions. I worked with other scholar-activists to expand discussions of globalization on our campuses and in the larger community. We observed and reported on the ways people were challenging globalization, and on how governments were responding to protests in increasingly repressive ways.

Many in the movement soon began speaking of the limits of "summit-hopping," and activists in Brazil and France called for people to come to Porto Alegre in Brazil to protest the World Economic Forum and put forward a new vision of globalization. Under the slogan "Another world is possible," they convened the first World Social Forum (WSF) in 2001, attracting around 20,000 activists. The next year they drew 60,000, and the following year 100,000. At the same time, hundreds of local, national, and regional social forums were organized to connect conversations across global and local levels and to expand possibilities for people to participate in the WSF (see Smith et al. 2012). Through my own participation in several World Social Forum meetings, European and U.S. Social Forums, and in local efforts in my community to relate to the World Social Forum process, I could see that activists were responding to a widespread desire to come together to share their experiences of economic globalization, develop analyses, and build collaborative networks based on a shared critique of economic globalization. More than 10 years later, the WSF continues to evolve and to inspire the imaginations of people seeking another world.

The two most concrete results of the WSF process are the *networks* it has helped initiate and sustain and the *ideas* it has helped develop and spread. Groups with diverse aims and constituencies—from peasant farmers to Indigenous activists to scholar-activists—have built long-term collaborative relationships through the WSF process. These networks, in turn, have helped generate new understandings and analyses, aiding the work of building broader alliances around more radically transformative demands. One prominent example is the emergence of a strong alliance supporting the call for food sovereignty which has been advanced by Via Campesina, a leading organization in the WSF process. Other examples are demands for global environmental justice responding to global climate change negotiations, and the various alternative or solidarity economies—such as community currencies and community supported agriculture—which allow communities to meet their needs outside globalized markets.

The WSF is thus an important space for doing sociological research and putting sociological tools into action. As I've attended different local, national and global forums and participated in organizing meetings of the U.S. Social Forum, I have learned how people develop long-term working relationships that enable coalitions to survive over time. I also can see through this kind of research the ways people who are embedded in local organizing networks think about and relate to global networks and politics. This has helped facilitate the work of translating ideas between global policy arenas and local communities and across different national and global contexts.

One particularly useful example is the idea of *intentionality* (Juris 2008). This idea recognizes that completely open spaces that prioritize participation can end up excluding many people according to race, class, and gender. So, in the U.S. Social Forum (USSF) process, my fellow USSF organizers have deliberately discussed how to ensure that leadership comes from those groups most harmed by global capitalism. This hasn't been easy, since people who have been marginalized by our economic system are not always able to come to meetings and to volunteer to do the organizing work—much less to take on leadership roles. In our meetings, we often need to take time to help newcomers learn about discussions we've had in the past and to facilitate participation by less experienced organizers. Intentionality means that more privileged activists are asked to step back to create space for these leaders and to work in solidarity to provide skills and resources that can support less privileged activists. The idea that the WSF is a *process* suggests that participants are learning new ways of working together and developing new kinds of relationships and identities. To learn how this process works and what its effects are, I have needed to engage in the participatory approaches I discussed above.

It is possible to identify some important accomplishments of the WSFs and the larger global justice movement which I have tried to support in my own work and with the resources and skills I bring as a scholar-activist. First, the movement has promoted a different story about globalization— one that stresses the negative consequences of global markets for both working and unemployed families around the world. These include, for instance, growing inequality which has limited the political voice of working people and furthered growing economic insecurity by lowering wages and increasing the *precarity* of work—that is, the lack of job security and decent working conditions. By obstructing global trade negotiations, the movement highlights the conflicts and inequities surrounding globalization. It denies elites the ability to claim that globalization is unambiguously good and universally beneficial. Throughout the world, more and more people have found it harder to make ends meet, despite the vast amount of wealth in the global economy. This story demands a change in how our economy is organized, and it encourages more to resist. The other important part of this story is that many thousands of people around the world are coming together to resist their conditions. This is not a story the commercial media wishes to tell, and even very large social forums have been ignored by mainstream press. I have tried to overcome this media blackout by telling this story through my teaching, research, and public engagement.

For me, being a sociologist in action has been a way to help support and encourage the WSF process and other efforts to build broad and diverse movements among people with very different experiences. Because we learn as sociologists to think about how larger social structures shape people's identities, perspectives, and interests, I find that I can often be of assistance in helping groups see areas of shared interests or possibilities for cooperative alliances. Also, as a teacher and writer, I have been fortunate to be able to help communicate movement ideas to larger audiences who might not otherwise hear them. Writing for popular media and online media sources has allowed me to perform a role that many activists lack the time and energy for, and my university credentials can lend legitimacy to the movement. The longer I have been involved in this work, the more I discover new insights about the dynamics of social change and the ways people in all places of society—including universities—must be involved if we are to realize a more just and peaceful world.

References

Juris, J. (2008). Spaces of intentionality: Race, class and horizontality at the United States Social Forum. *Mobilization 13*, 353-372.

Santos, B. (2004). The World Social Forum: Toward a counter-hegemonic globalisation (Part I). In J. Sen, A. Anand, A. Escobar, & P. Waterman (Eds.), *The World Social Forum: Challenging empires* (pp. 235–246), Montreal, QC: Black Rose Books.

Smith, J. (2001). Globalizing resistance: The Battle of Seattle and the future of social movements. *Mobilization 6*, 1-20.

Smith, J., & Moran, T. P. (2000). WTO 101: Myths about the World Trade Organization. *Dissent.* Retrieved from http://dissentmagazine.org/article/?article=1498

Smith, J., Byrd, S., Reese, E., & Smythe, E. (Eds.). (2011). *Handbook of World Social Forum activism.* Boulder, CO: Paradigm Publishers.

Wallach, L. & Woodall, P. (2004). *Whose trade organization?: A comprehensive guide to the WTO.* New York, NY: The New Press.

INFORMED DEBATE IN A POLITICAL MINEFIELD

Irene Bloemraad

University of California, Berkeley

Irene Bloemraad is associate professor in sociology at the University of California, Berkeley, and a scholar of the Canadian Institute for Advanced Research. Her research focuses on immigration and politics, including citizenship, immigrants' political and civic participation, and multiculturalism policies. Her publications include *Becoming a Citizen* (2006), *Civic Hopes and Political Realities* (edited with Karthick Ramakrishnan, 2008) and *Rallying for Immigrant Rights* (edited with Kim Voss, 2011). Professor Bloemraad also enjoys speaking to community groups, policymakers and the general public about immigration.

I had been speaking for about 45 minutes when a hand went up in the back. The hand did not belong to a student, and I wasn't standing in a lecture hall, but rather in the conference room of the Livermore, California, police station. I and the 60 other people in the room had traded a lazy Saturday morning for a discussion about immigration. Congress was debating immigration legislation, including possible legal status for migrants living in the United States without proper documentation, so the Livermore-Amador Valley League of Women Voters had invited me to speak at their annual meeting.

My audience was overwhelmingly female, older, and white, many the descendants of European immigrants who had come to the United States a century earlier. Northern California is known for progressive politics, but a number of the audience members felt that there were problems with "today's immigrants," whom they believed were not learning English as quickly nor integrating as fast as their own parents and grandparents did. As one gentleman put it, "My grandfather told all his kids they needed to learn English. You only spoke Yiddish at home."

I thought that the owner of the raised hand, a woman in her mid-40s whose brown-hued skin and dark hair suggested Latino origins, would speak favorably about today's immigrants. I was a bit surprised when she said, "I want to know how we can stop all these illegals. My husband and I own a construction company and we're being put out of business by everyone hiring illegals." She went on to explain that she was proud of her Mexican American heritage, but the government needed to protect Americans' economic interests first.

I shouldn't have been surprised by the woman's question. As I regularly tell my students, the politics of immigration raises concerns for many Americans. Policy preferences cut across traditional political lines. A sizeable number of well-meaning Americans fear that migrants drive down wages or are changing the character of the country too rapidly; in 2009, a major survey found that 54% of U.S. respondents said immigration was more of a problem than an opportunity.[3] In contrast, big business often favors migration as a source of workers; the U.S. Chamber of Commerce, for example, supports the idea of amnesty for undocumented migrants. Having concerns about immigration doesn't immediately make someone racist, and supporting amnesty isn't automatically a sign of unselfish humanitarianism.

This reality—that public concerns and social science evidence about immigration are not simple matters of good or bad, right or wrong—is a key reason why I give talks to community groups. The politics of immigration quickly becomes heated, with advocacy and interest groups, politicians and pundits all throwing around "facts" in support of their preferred position. As a sociologist, I try to present a fuller range of facts. I let people know what social scientists have found out about immigrants' integration into U.S. society and the effects of migration on the economy. I also try to give them tools to make their own informed choices about policy options and public debates, to go beyond the emotional appeals and alarmist messages that make immigration such a political minefield.[4]

Especially in California, but increasingly throughout the United States, the question of how the United States should deal with undocumented migrants tops Americans' concerns about immigration. The stakes in

---

[3]This survey, funded by a group of nonpartisan foundations in the United States, Canada, and European countries, found that Americans ranked third out of eight countries in their concern over migration. Concern was highest in the United Kingdom, where 66% of respondents felt immigration was more a problem than opportunity, and lowest in Canada, where only 25% agreed with the sentiment. The report, *Transatlantic Trends: Immigration, 2009*, can be found at www.transatlantictrends.org, last accessed on December 14, 2012.

[4]On the question of linguistic integration, for example, I showed the Livermore group a graph of English and Spanish language ability among those who were born in Mexico, the children of Mexican immigrants, and their grandchildren. Those who move to the United States as adults speak primarily Spanish, but their children almost all speak English fluently, and a fair number are bilingual. Among the grandchildren of Mexican immigrants, almost no one speaks Spanish fluently. The scientific evidence showing linguistic assimilation is overwhelming. I then raised a normative question: Should we view assimilation to monolingual English as a good thing, an example of successful integration? Or should we see the loss of Spanish in the second and third generations as a missed opportunity for cultural enrichment and potentially valuable linguistic skills for a globalizing economy?

answering this question are large: About 11.5 million people live in the United States without legal residence papers. Many of them have U.S.-born children and other relatives, so what happens to them affects entire families.

I took my audience through the main policy responses under debate: (1) tighter border control, (2) finding and deporting unauthorized migrants, (3) penalizing employers, (4) legal regularization (amnesty, temporary visas), and (5) foreign aid to help develop the economies of countries of emigration so that fewer people will want to migrate to the United States. On border control, I showed a graph: If you plot the number of undocumented migrants in the United States from 1980 to the present, you see dramatic increases in the number of undocumented people *after* each increase in border control efforts. I asked my audience to become social scientists: What is going on? Isn't border control supposed to *halt* undocumented migration?

A few people smiled at the invitation to become a social scientist, but everyone was intrigued and eager to figure out what is going on. After an animated discussion, a plausible story emerged. A significant proportion of the undocumented population in the 1970s and 1980s were men who worked in agriculture, construction, and other seasonal employment. They engaged in *circular migration*—moving to the United States to work at certain times of the year, and then heading back home to family and friends at other times. Erect a wall—real, or in the form of border guards and surveillance technology—and going back becomes difficult. Some of those left behind decided to reunite with their loved ones in places like California by also migrating. We find a classic case of unintended consequences: The very tactics used to resolve the problem might instead exacerbate it.[5]

What I do during community presentations resembles what goes on in my classroom when I teach on the Berkeley campus. I share information with my audience members, I push them to think through evidence that might support or undermine a proposition, and I engage in discussion so that those listening are not just passive spectators, sitting in front of some sort of academic television, but rather are engaged in the process of learning and evaluation.

---

[5]This is not the only explanation for the rise in the number of undocumented migrants, but it likely accounts for the largest significant part of the increase. Other explanations include the implementation of the North American Free Trade Agreement, which has affected the Mexican economy; recruitment by American businesses seeking out migrant labor; the U.S. government's refusal to accept the asylum claims of certain Central American migrants; and growth in the number of people who overstay the time limit on their temporary student, tourist, or work visas.

Depending on the audience, I might put more emphasis on providing information or teaching people how to collect their own information. In Livermore, I mostly provided information. Another time, when I spoke at a training session for lawyers and immigration service providers organized by CLINIC, the Catholic Legal Aid Network, I focused on giving participants research tools.

Those at the CLINIC training session always struggle to fund their nonprofit organizations, which provide assistance to immigrants. Grants, offered by foundations or organizations like the United Way, provide one source of funds, but to receive these grants, community organizations must put together competitive and comprehensive proposals explaining why they need the money and what they will do with it. Unfortunately, some of those working in this field lack the research training to do this easily, and most don't have the time to engage in extensive studies. So I showed participants some easy ways to assemble statistics using available resources, such as the U.S. Census Bureau Web site; information from the California Department of Education; and online data and reports from the Migration Policy Institute, a Washington, DC–based think tank. My presentation was a bit like what I do when I teach research methods to sociology majors at Berkeley, but in this case, I was giving people tools to support their advocacy and service activities with immigrants.

Teaching and research—which is the process of using evidence to adjudicate questions—are the core skills of an academic sociologist. In my volunteer time, I try to bring those skills outside the university classroom. Why should college students be the only ones to puzzle through questions, assemble evidence, and debate conclusions? I don't tell people whom to vote for or what policy stance to take on debates about immigration, but I do push them to become intelligent and informed citizens, using social science evidence to weigh options and consider alternatives. I was gratified when, after my presentation in the Livermore police station, the woman who had asked about illegal migration came up to chat with me. "I hadn't really considered it that way," she said. "You've really got me thinking. Thank you."

Before entering a graduate program to earn a PhD in sociology, I considered working in government as a policy analyst. I felt that this would be a great way to marry the research and analytical skills I learned in school with my wish to have a direct impact on people's lives. I ended up taking an academic path instead, but speaking to community groups allows me, in a small way, to use my sociological tools to affect people's lives outside the classroom, and perhaps improve the policy decisions that come from informed public debate.

COMMUNITY-BASED RESEARCH AND IMMIGRANT RIGHTS

Leah Schmalzbauer

Montana State University, Bozeman

Leah Schmalzbauer received her PhD from Boston College and is an associate professor in the Department of Sociology and Anthropology at Montana State University. An activist scholar, Leah has won several teaching and research awards. She is currently completing a five-year ethnographic project exploring gender relations and family formation among Mexican migrants in the rural Mountain West. Leah will be a visiting scholar at the Center for Social Anthropological Research and Exploration (CIESAS) in Oaxaca, Mexico, during the 2012-2013 academic year. There she will complete a book, currently titled *The Last Best Place?: Gender, Family and Migration in the New West*. Leah's first book, *Striving and Surviving: A Daily Life Analysis of Honduran Transnational Families*, was published by Routledge Press in 2005.

On April 6, 2008, about 100 undocumented Mexican migrants who live in Bozeman, Montana, gathered at Resurrection Catholic Church for a Know Your Rights forum. An immigration lawyer led the forum, educating attendees about their rights, and advising them about what to do if these rights are violated. The lawyer did not speak Spanish. Indeed, there are still no immigration lawyers in Montana who do. Therefore, Bridget Kevane, a Spanish professor at Montana State University, interpreted. I sat on the side of the sanctuary with my newborn in my arms, amazed by the intensity of the conversation. Carlos, a young man from Zacatecas, Mexico, asked, "What do I do if I am pulled over and the police officer asks for my immigration papers? I know they don't have the right to ask, but what if they do?" The lawyer repeated the answer that she had given several times that day,

> You need to just stay quiet. The police in Montana do not have the right to act as Immigration and Customs Enforcement [ICE] agents, and, therefore, they do not have the right to ask for your papers. As difficult as it is, you need to just stay quiet.

The lawyer proceeded to pass out half-sheets of paper that stated in basic legalese the rights of all immigrants in the case of police apprehensions. She advised all attendees to keep the statement with them at all times, and to hand it, without speaking, to the police if they are ever apprehended.

With a grant from the American Sociological Association Community Action Research Program, Professor Bethany Letiecq (Department of Health and Human Development at Montana State) and I organized this Know Your Rights forum as a response to the abundant data I had gathered in two years of field research documenting a "culture of fear" among Mexican migrants in Montana (see Schmalzbauer, 2009). While this fear has many roots, one of the most prominent is the knowledge that police stops are what most commonly lead to deportations in our state. Because Southwest Montana has such a short history of Latino migration, there are few formal supports available for migrants. In Bozeman, for example, the hub of Montana's new wave of migration, there is not a single organization dedicated to immigrant rights. In addition, there are few people in the area who speak Spanish and fewer still who have even an elementary knowledge about immigration issues. As an immigration scholar who speaks Spanish, I have thus found many opportunities to bridge my academic research with local immigrant rights work. The Know Your Rights forum is just one example of this bridging.

The context of migration in Montana is complex. Indeed, there are two parallel, yet very different, migrations changing the demographics of the state. The first migration is that of wealthy retirees and telecommuters from the West Coast, primarily California, who are building and refurbishing expensive homes and ranches in the region. This influx of wealth has spurred the demand for low-wage workers to construct new homes and mountain resorts. This demand has been filled by Mexican migrants, primarily from California, Colorado, and Idaho, who have been recruited by local employers to work in the booming construction industry. This demographic shift has sparked many questions for me: What is it like for Mexicans to migrate to a small, predominantly white town in the Rocky Mountains, where few people speak Spanish, snow can fall seven months of the year, and housing costs far surpass the national average? How, literally, are they surviving? Many scholars, myself included, have researched the experience of Latino migrants in U.S. cities (see Hondagneu-Sotelo, 2001; Levitt, 2001; Schmalzbauer, 2005; Smith, 2006), but we still know little about the lives of migrants in rural, nontraditional destinations of the United States (Schmalzbauer, 2009).

I began my field research in Montana as a participant-observer (see Hesse-Biber & Leavy, 2006), translating at the local food bank, one of the few "safe" spaces in town for migrants. Through my work there, I met many Mexican women and began to hear their stories. I also became an informal social worker of sorts, translating at doctor's appointments and the public assistance office, and helping migrants negotiate the school system. In the process, I became intimately involved in certain aspects of their lives. After months of participant observation, I did my first formal in-depth interview to further explore the daily survival struggles and strategies of migrants in Montana. From there, my interview base grew quickly. The social capital that I had accumulated through my community work gave me the privilege to record many life stories, providing rich data for academic writing. The relationships I have built with community members have also confirmed my commitment to doing research with a social justice component, opening my eyes to the many needs, struggles, and strengths of Latinos in Montana.

My research has always been a collaborative pursuit. When I was in Boston conducting my dissertation research on Honduran transnational families, I partnered with Proyecto Hondureño, an immigrant rights organization there. Leaders of that organization gave me guidance on every phase of my research and were present at my dissertation defense. They later used the book (Schmalzbauer, 2005) that materialized from my dissertation to gain support from funders and as fuel in their political efforts. Although I am no longer in Boston, I remain closely connected to the organization and to the Honduran community there.

I have followed a similar collaborative path in Montana. Early on in my field research, I sought out community partners. I began by contacting Professor Letiecq, who is the chair of the Gallatin Valley Human Rights Task Force, to ask her advice about where migrants could turn if they were not being paid by their employers, if they were arrested and did not have legal representation, or if they were afraid to apply for food stamps—all issues that had surfaced repeatedly in my field work. Professor Letiecq in turn contacted Kim Abbott, a field organizer for the Montana Human Rights Network. The three of us sat down in the fall of 2007 to talk about the struggles facing migrants in Montana and to strategize about what we could do to help ease them. Much was born from this meeting. Most importantly, the Montana Human Rights Network adopted immigrant rights as one of their primary platforms, and Professor Letiecq and I partnered to do a community-based research project to document, analyze, and address the most urgent needs of the Montana Mexican community.

The Know Your Rights forum with which I opened this piece is but one of the activities that have materialized from my collaborative research efforts. In partnership with Professor Letiecq and Kim Abbott, I have also organized legal clinics where migrants have the chance to sit down, one-on-one, with an immigration lawyer to discuss employment, housing, or their immigration status. I have organized a public educational panel on Mexican migration, and with Letiecq and Abbott have contributed testimony and data to help defeat anti-immigration legislation. I have also supported local cultural events to bridge the Latino and native-born communities. The most notable legislation that we organized to defeat would have deputized local police officers to act as ICE agents. In legislative sessions on the issue, we shared narratives of the fear and anxiety that police stops have sparked among local Latinos, emphasizing their subsequent hesitancy to report crimes and to trust law enforcement. We have also been active in educating our Montana legislators and interested public about comprehensive immigration reform as well as organizing support to include immigrants in health reform legislation. Professor Letiecq and I have also formed a community advisory board, composed of five Mexican migrants, to assist us in our research efforts. We meet once a month to hone our research questions, check our interview guides, and analyze our data. With grant money from the National Institute of Health Montana INBRE (IDeA Network for Biomedical Research Excellence) program, we have also been able to train local Mexican community members to help us carry out our research.

I am confident that my community efforts have strengthened my academic scholarship and contributed to the creation of a more just and welcoming community. On the academic level, the relationships that I have built with Mexican migrants have given me access to rich and complex data that would have been impossible to gather using traditional methods that lack a participatory component. As a result, I have been able to contribute new knowledge to the sociology of immigration, as well as to infuse my teaching with my research on the local migration context. The community advisory board with whom Professor Letiecq and I are working has given us unique insight into our data and guidance in the field that the immigration literature alone could not have provided. We have all been empowered through the collaborative process.

On the community level, I have seen solidarity and leadership emerge within the local Mexican community easing, even if only slightly, the isolation, marginalization, and fear that so many migrants experience here on a daily basis. In the last meeting of our community advisory board, Dora, an undocumented single mother who has struggled mightily to survive in Montana, told our group "For the first time since arriving in Montana, I feel like we are becoming a part of the local community." Her comment followed a Mexican dinner and dance that we helped organize as part of Mexican Education Week at Montana State University, which brought together many Mexicans and white community members in a wonderful celebration. Students who have studied migration with me have also played a major part in easing community isolation, teaching English to interested migrants for service-learning credit. We in Southwest Montana still have a long, long way to go to achieve security, opportunity, and justice for all community members, but I feel that the collaborative research, teaching, and organizing efforts of which I have been a part are making a difference.

References

Hesse-Biber, S., & Leavy, P. (2006). *The practice of qualitative research*. Thousand Oaks, CA: Sage.

Hondagneu-Sotelo, P. (2001). *Domestica: Immigrant workers cleaning and caring in the shadows of affluence*. Berkeley: University of California Press.

Levitt, P. (2001). *The transnational villagers*. Berkeley: University of California Press.

Schmalzbauer, L. (2005). *Striving and surviving: A daily life analysis of Honduran transnational families*. New York, NY: Routledge.

Schmalzbauer, L. (2009). Gender on a new frontier: Mexican migration in the rural Mountain West. *Gender & Society, 23*(6), 747–767.

Smith, R. C. (2006). *Mexican New York: Transnational lives of new immigrants*. Berkeley: University of California Press.

DISCUSSION QUESTIONS

1. Ellie Benard was able to make a positive impact in her community while carrying out a research project as an undergraduate. How do you think this opportunity impacted (1) how she views her ability to make a positive impact on society and (2) her career goals? Can you think of ways you might be able to conduct research that can similarly impact your community?

2. Bandy and Benard speak of tensions between Somali immigrants and native Mainers. Imagine that you are a Sociologist in Action called in to explain and help mitigate tensions between longtime residents and recent immigrants in your town. What sociological information would you need to know before you could begin to figure out how to address the situation?

3. How does Becky Hsu's research reveal that cultural norms must be considered when establishing a microcredit program? How does her work illustrate the importance of understanding how a society operates before trying to improve it?

4. If you were going to set up a microcredit program for people where you grew up, do you think it would be successful? Why or why not? What are some of the cultural norms that would impact its success? Do you think that there are people in the community where your college is located who could benefit from a microcredit program? If so, how might students go about working with your college and leaders of the local community to create such a program?

5. Jackie Smith describes how, when she was growing up in the 1980s, adults in her church and her school introduced her to international peace and justice activism. Did you have a similar experience when you were growing up? If so, please describe it and its impact on your personal and career goals. If not, how do you think having had such socializing agents as a child would have influenced your personal and career goals?

6. Smith describes how the idea of *intentionality* "recognizes that completely open spaces that prioritize participation can end up excluding many people according to race, class, and gender." Explain how and why this can happen and what sociological tools might be used to make sure that more privileged participants make space for others to take on leadership roles in an organization.

7. Irene Bloemraad takes her teaching outside the classroom to a variety of public audiences. Do you think it's important for citizens to be exposed to the sociological perspective when debating and voting on such topics as immigration? If so, why? Have you conducted any sociology research projects where you have become informed on a topic you could take out to public audiences (e.g., holding a forum on campus, speaking to local junior high or high-school students, etc.)? If yes, how might you go about setting up avenues to bring your research (and that of other students) to these public audiences? If no, what topic would you want to learn more about in order to better inform others?

8. According to Bloemraad, why does increased border control lead to higher numbers of undocumented immigrants living in the United States? How might you address this issue and the information gap about it?

9. How was Leah Schmalzbauer able to gain the trust of the Mexican immigrants she studied? Why is it important for (1) ethical and (2) data collection reasons that researchers make their agendas clear to the people they are studying?

10. Would you be interested in studying an immigrant community in the United States? If no, why not (and what is another group you would be interested in examining—and why?). If yes, which group? Why? What are the top three interview questions you might ask? Why?

11. How might Hsu's findings on the importance of culture further the work of Schmalzbauer and Benard with immigrants?

12. Which of the Sociologist in Action pieces in this chapter did you find most interesting? Why? If you had the opportunity to work with one of them, whom would you choose? Why?

RESOURCES

The following Web sites will help you to further explore the topics discussed in this chapter:

BBC News "Factfile: Global Migration"	http://news.bbc.co.uk/2/shared/spl/hi/world/04/migration/html/migration_boom.stm
Global Exchange	http://globalexchange.org/
Immigration Myths and Facts—January 2008	http://www.aclu.org/files/pdfs/immigrants/myths_facts_jan2008.pdf#page=1
International Forum on Globalization	http://www.ifg.org/
Public Citizen Global Trade Watch	http://www.citizen.org/trade
The Miniature Earth	http://www.youtube.com/watch?v=fA6MhyK60iI

To find more resources on the topics covered in this chapter, please go to the Sociologists in Action Web site at **www.sagepub.com/korgensia2e.**

Chapter 12

Environmental Justice

Environmental justice refers to (1) efforts to ensure that environmental quality and hazards are consistent across social classes, races, ethnicities, and regions, and (2) efforts to ensure that human beings operate in sustainable ways. Sociologists who work for environmental justice look for patterns of environmental inequality, seek to understand why such patterns exist, and strive to use their sociological knowledge to alleviate them. They also work to alleviate such dangers to our society as global warming and other socially created environmental hazards. The Sociologists in Action in this chapter reveal how their use of sociological tools has propelled them to the forefront of the environmental justice movement.

Andrea Rother begins this chapter with "Reducing Pesticide Exposure Risks: An Environmental Sociologist's Role" describing how she uses her training as an environmental sociologist to recognize and understand cultural differences in order to advocate for the safer use of pesticides. Her work has enabled Kenyan farmers to implement pesticide-alternative techniques, improved the clarity of pesticide labels, and helped remove dangerous, unlabeled pesticides from the streets in and around Cape Town, South Africa. In the process, she has saved countless lives.

Lou Jacobson shares how he uses sociological tools to help mitigate global warming through promoting energy efficiency in "Using a Sociological Tool Kit to Make Energy Efficiency Happen." Growing up in the heart of the Appalachian Mountains, Jacobson saw firsthand how our reliance on coal for electricity has resulted in mountaintop removal, poisoned streams, and other environmental degradation. In his work with the Redwood Coast Energy Authority, Jacobson uses the sociological knowledge he gained from his master's program in sociology. His sociological tools help him to understand why some people might be hesitant to take steps to reduce their energy consumption, to effectively address their concerns, and to enable them to conserve and use energy more wisely. His efforts and those of the people with whom he has consulted benefit all of us.

David Pellow has led impressive environmental justice efforts all over the world. In "Activist Scholarship for Environmental Justice," he discusses his work organizing and working with the Transatlantic Initiative on Environmental Justice (TIEJ) and other organizations to address such environmental injustices as a Roma refugee camp being located on a toxic waste site in Kosovo. He also describes the impact of a guide he and others created to help "ordinary people [across the globe] confront large mining companies that attempt to take control of communities to ensure profits for their shareholders no matter the cost to the local people and ecosystems." Pellow's efforts illustrate that environmental injustices are socially created and can be effectively addressed by those ready and willing to use sociological tools.

Daniel Faber concludes this chapter with "The Sociology of Environmental Justice: Merging Research and Action." His work uncovering dramatic environmental inequality in Massachusetts caused "shockwaves" throughout the state. Faber and his colleague revealed that American environmental injustices that lead to lower-income families and people of color facing much greater health risks don't happen just in the Sunbelt or other areas of the United States known for relatively corporate-friendly environmental policies. In this piece, he describes how he studies "the disparate exposures to environmental health hazards experienced by people of color and working-class whites," and how he and others use sociological tools to recognize these patterns and to organize effectively to address them. Faber recognizes that "environmental activism can be effective only when guided by sociological prescriptions that take into account the interconnections between social and ecological issues."

REDUCING PESTICIDE EXPOSURE RISKS: AN ENVIRONMENTAL SOCIOLOGIST'S ROLE

(Hanna-) Andrea Rother

University of Cape Town, Cape Town, South Africa

Andrea Rother, PhD, is head of the Health Risk Management Programme in the University of Cape Town's (UCT) School of Public Health and Family Medicine. In 2010, she received UCT's prestigious Vice Chancellor's Award for her social responsiveness work on identifying the link between street pesticides and child poisonings. In 2011, she established a new postgraduate diploma in Pesticide Risk Management (unique in incorporating sociology and science) in conjunction with the United Nations (UN) Food and Agricultural Organization (FAO). She serves as a pesticide/chemical and risk communication expert for several UN organizations including the World Health Organization (WHO), Food and Agriculture Organization (FAO), Environmental Programme (UNEP) and Institute for Training and Research (UNITAR).

A question I am often asked by students is "how did you get into the field of pesticides?" I first started defining myself as an environmental sociologist specializing in pesticides specifically—and chemicals more generally—as a master's student in rural sociology at Michigan State University (MSU). I had spent a year at the University of Zimbabwe through an exchange program and returned to MSU to study ChiShona (a Zimbabwean language). As I spent hours learning agricultural phrases in ChiShona, so I could conduct PhD research in Zimbabwe, I researched the role pesticides were playing in the demise of traditional African farming practices.

In order to understand the sociological implications of pesticide use, I took a pesticide chemistry and ecotoxicology course and read books that covered social-technological conflicts[1]. The power struggle between profit-driven industries and protecting human health and the environment drew me into the world of pesticides—an environmental sociologist's playground!

In 1994, I received funding to conduct preliminary research for my PhD dissertation in Kenya and Zimbabwe. My experience riding around in a 4x4 truck with an all-male research team, visiting farmers in remote and stunning areas of southern Kenya, revealed that the work of the team lacked the essence of sociological research. The researchers were focusing on introducing pesticide alternative techniques and hybrid plants, rather than listening to why or why not the farmers would adopt these alternative methods and plants.

My extensive research training in sociology and anthropology led me to conduct informal interviews with the farmers asking only what they liked and disliked about these methods. The result of six weeks of listening to farmers and walking around their fields was (1) my discovery that the farmers had misinterpreted the project's intention and what was expected of

---

[1]Such books include *Silent Spring* (Carson, 1962), *Circle of Poison* (Weir & Schapiro, 1981), *Pesticides and Politics* (Bosso, 1987), and *The Death of Ramon Gonzalez* (Wright, 1992).

them and (2) the salvation of the project. Once I explained to the research team that the farmers did not understand the point of the project, the team quickly held clarification meetings with the farmers and produced a booklet detailing what they wanted the farmers to do. Afterwards, the farmers held a village meeting where they awarded me a wooden carved lion as a sign of their appreciation for my taking the time to listen to their opinions.

This experience in Kenya opened my eyes and confirmed the important role sociologists play in natural science fields. I kept that in mind as I conducted research about the ways in which Zimbabwean vegetable farmers understand the risk information printed on pesticide labels. The results illustrated that their *risk consciousness* (Beck, 1992; Jensen & Bok, 2008) and risk perceptions (i.e., beliefs and attitudes about risks) led them to design their own protective equipment (e.g., raincoats and head scarves for face masks) and grow vegetables without pesticides for their household consumption.

Risks, such as those that derive from pesticide use, are not simply objective conditions "out there" waiting to be perceived by individuals, but are *social constructs* (Dake, 1992). That is, what is perceived to be a risk is based on individuals' beliefs and understanding of *risk*, and these are influenced by their social and cultural frame of reference. I began to question how farm workers, small-scale farmers and the general public—particularly in developing countries—are meant to find out about and understand the technological risks from pesticides that are determined through scientific experiments and risk assessments.

My PhD research with farm workers in South Africa assessed the extent of risk perceptions' influence on how the information on the pesticide label is understood, and in turn how the pesticide label as a risk communication tool influences people's risk perceptions. The legal status of pesticide labels is twofold—one is that specific information must be on the label of pesticides sold, and the second is that people who use pesticides have a legally binding responsibility to read, understand, and follow the label instructions. I was incensed. How are people with low literacy levels, literacy in another language, or with no formal education to understand the risk information on a label?!

Even interpretations of *pictures* used to give warnings of hazardous material that label designers deem universally understandable can vary from culture to culture and be misinterpreted. Pictograms are not intuitively obvious. To illustrate my point, what do you think the following pictogram currently being introduced onto chemical labels globally means, and what action would you take in response to the pictogram? ⬦[2]

---

[2]This symbol means that the chemical is potentially a chronic hazard. That is, prolonged or repeated exposure to the chemical may cause long term health effects, such as cancer or birth defects.

Now look at footnote two. Did you get it right? Do you think pictograms can cross-culturally transmit risk information without words and provide what safety action must be taken? Did everyone in your class have the same understanding or interpretation of this pictogram?

As a result of my research, I developed several risk communication tools (e.g., label cards and stickers: see http://www.coehr.uct.ac.za/publications/pestrel.php) so that farm workers and the general public can have access to what the information on pesticide labels means. What is not ensured is that the risk information is comprehended as scientifically intended. Even if the information is understood, there is no way to ensure that the individual will then evoke a safety behavior to prevent harmful exposures. Risk communication tools can go a long way to assist with efforts to ensure that symbols, icons, and colors are interpreted as intended, and sociologists can be instrumental in developing these tools through research with the target audience.

In 2006, I was approached by a local Cape Town nongovernmental organization (NGO) requesting assistance in dealing with informal vendors selling street pesticides. The occupational hazards were high and the NGO sought my assistance to help find a way to maintain the economic livelihood of the vendors while reducing harmful exposures to sellers and consumers. Most people assume that pesticides are predominately used in agriculture. However, informal sellers are putting agricultural pesticides (too toxic for domestic use) into empty water, juice, and liquor bottles (sometimes water is added) and selling them for the control of cockroaches, flies, fleas, and bed bugs. Some sellers peddle Aldicarb, considered one of the most acutely toxic agricultural pesticides and banned in the United States, in small plastic strips for use on rats. The locals call it "Two Step," which is all the rats can take before dying. As a result, a young child eight months old (who sucked on a strip that had once held the pesticide) was poisoned and admitted into a local children's hospital. It is difficult to link poisonings to unlabeled street pesticides, but through collecting narratives and reading case files, we identified that 68% of suspected poisoning cases during 2008–2009 were from street pesticides.

The challenge was to prove that people, especially children, were being poisoned by these products because they are unlabeled. In order to do this, I used several methods (Rother, 2010a). First, I had research assistants travel the trains and go to the local markets to inquire about and buy the products. They then took the products to a laboratory to identify the active ingredients (The lab refused to do any more samples after 10, as they complained the high toxicity ruined and contaminated their machines!). I had students visit a local children's hospital every time poisoning cases were admitted to take the narratives of their story so we could try and link the cases to street pesticides based on descriptions of the product and where it was bought. After collecting this information and conducting household surveys in two townships, where we handed out 400 rat traps and assessed their acceptance by community members, we began advocacy work.

We developed and distributed several risk communication tools, translated into local languages. I set up a Child Pesticide Poisoning Reference Group with government officials and researchers, through which we discussed several intervention strategies. The Health and Agriculture Departments are now removing these pesticides from the streets (Rother, 2010b). We also developed an algorithm on how to identify street pesticide poisonings and to report these—20,000 copies of these laminated cards were distributed to local clinics and hospitals.

The hardest part of my research was standing over the bed of a two-year-old, poisoned by accidently drinking a street pesticide, and interviewing a young woman who bought Aldicarb on the street and ate it to commit suicide. These incidents, however, just further fuel my desire to use my skills as a sociologist to prevent pesticide poisonings locally and globally. Without my training as an environmental sociologist, I would not be able to play an effective role in poison prevention and highlighting exposure risks. As a sociologist in action, I know that I am not just working through the issues on paper, but actually out in communities and working with policy makers to affect change, as well as inspiring the next generation to pick up the pesticide justice torch and run with it!

References

Beck, U. (1992). *Risk society: Towards a new modernity* (English trans.). London, UK: Sage.

Boso, C. (1987). *Pesticides and politics: The life cycle of a public issue.* Pittsburgh, PA: The University of Pittsburgh Press.

Carson, R. (1962). *Silent Spring.* London, UK: Hamish Hamilton.

Dake, K. (1992). Myths of nature: Culture and the social construction of risk. *Journal of Social Issues, 48*(4), 21–38.

Jensen, M., & Blok, A. (2008). Pesticides in the risk society: The view from everyday life. *Current Sociology 56*(5), 757–778.

Leiss, W., & Powell, D. (2004). *Mad cows and mother's milk—The perils of poor risk communication* (2nd ed.). Montreal, QC: McGill-Queen's University Press.

Rother, H-A. (2008). South African farm workers' interpretation of risk assessment data expressed as pictograms on pesticide labels. *Environmental Research 108*(3), 419–427.

Rother, H-A. (2010a). Falling through the regulatory cracks: Street selling of pesticides and poisoning among urban youth in South Africa. *International Journal of Occupational and Environmental Health 16*(2),202–213.

Rother, H-A. (2010b). Poverty, pests and pesticides sold on South Africa's streets. In J. Lee and S. Shaw, *Women worldwide: Translational feminist perspectives on women* (1st ed.), No.82. New York, NY: McGraw-Hill.

Rother, H-A. (2011). Challenges in pesticide risk communication. In Nriagu JO (Ed.), *Encyclopedia of Environmental Health 1*, 566–575.

Weir, D., & Schapiro, M. (1981). *Circle of poison—Pesticides and people in a hungry world.* San Francisco, CA: Institute for Food and Development Policy.

Wright, A. (1992). *The death of Ramon Gonzalez—A modern agricultural dilemma.* Austin: University of Texas Press.

USING A SOCIOLOGICAL TOOL KIT TO MAKE ENERGY EFFICIENCY HAPPEN

Lou Jacobson

Redwood Coast Energy Authority

Growing up in southeastern Ohio, Lou Jacobson was exposed to mountaintop removal, clear-cutting, and illegal trash dumps. His childhood, combined with experiences in early adulthood, nurtured an interest in understanding and serving society while protecting the environment. Soon after his undergraduate graduation, he began his life as a sociologist in action. He worked with a West Virginia whitewater rafting company as a guide and to benchmark and reduce consumption. Currently, Lou Jacobson's desire to serve society and protect the environment is expressed in his work mitigating climate change and environmental degradation through energy efficiency.

Living in the heart of the Appalachian Mountains shaped my decision to study energy efficiency during my graduate studies in sociology. I saw, firsthand, the environmental damage caused by coal-generated electricity. Over the years, I watched mountains that once stood high disappear because of society's "need" for cheap electricity. I researched what I could do to help and realized that I could use less electricity by making a few small changes at home like using power strips to reduce phantom (vampire) loads, turning down my thermostat, weatherizing my home, and using energy-saving compact fluorescent light (CFL) bulbs.

In conducting my research, I realized that energy efficiency could drastically help mitigate climate change while limiting mountaintop removal.

Unfortunately, my initial research indicated that changes such as the ones I had made at home were not rapidly being embraced by individuals across society. I therefore decided to focus my graduate work on understanding why this was the case.

My graduate studies in sociology provided me with a set of tools that I use in my work to mitigate climate change. Graduate classes and readings that covered *diffusion theory* (how new technologies or ideas are accepted and adopted across communities) and *risk perception* (the subjective perception of the extent and severity of risk) influenced how I currently discuss and act toward energy efficiency opportunities. Seminars in research methodology honed my data collection and interpretation skills, allowing me to determine how to most effectively interact with energy users and avoid stereotyping them. Finally, carrying out my master's thesis on the barriers to the diffusion of the CFLs helped me gain specific and useful knowledge for my present career as an energy efficiency expert.

Shortly before I earned my master's in sociology, I was hired at the Redwood Coast Energy Authority (RCEA), which is a Joint Powers Authority in Humboldt County, California. RCEA partners with the utility company to reduce power demand and electricity consumption by offering a variety of commercial and residential community services. The partnership is funded under the auspices of the California Public Utilities Commission (CPUC) and is administered by the utility company.

As an energy specialist at RCEA, it is my job to serve the community's commercial energy efficiency needs. To do this, I provide a variety of services including free, energy-efficiency assessments and project management. During the assessments, I look for and note technological or behavioral energy-saving opportunities. After the assessment, I research and propose energy-efficient solutions. I include, in a report, the incentives associated with technological changes and any other pertinent behavioral information relating to the original audit. When the report is complete, I return to the business and present the information. Should the company opt in to a technological change, I facilitate the process by managing the project, reporting the work to the local utility so they can claim energy savings with the CPUC, and issuing the incentives to the installing contractor after the technologies are updated and inspected.

I've found that significant financial incentives and a better long-term bottom line are very important. However, they don't always fully drive businesses' energy efficiency decisions. Often other social issues present themselves as barriers to energy efficiency. In these cases, I have the latitude to address each decision maker based on his or her unique situation. Although I understand, in general, the different social barriers to energy efficiency, I am careful not to stereotype individuals. Each situation is different, and it's never as easy as asking, "What can I do to get you to change?" Here's when my sociological training comes in to play! I try to closely observe and note each interaction with my clients. I attempt to ask questions to better understand how they see the situation, and I have developed a specialized database to keep track of the observations. What I've found is that by applying sociological tools like diffusion theory, social theories of risk perception, previous literature on the social barriers to energy efficiency, and research methodologies, I have a better chance of identifying and overcoming potential barriers to energy efficiency.

For example, I had one local nonprofit initially refuse a project that I could fully fund and manage. The executive director (ED) thought it was a scam even after I delivered all the information indicating that the money came from local charitable foundations and state funds. He kept on asking me what the catch was, so I decided that I needed to take a different approach. After reviewing my field notes several times, I noticed a recurring theme. Several comments the ED had made suggested that he was skeptical of governments and corporations. Since I work for a local government organization in partnership with a corporation, I interpreted this as an indication of why he might not believe me or the clear information I presented.

Diffusion theory framed my next step by suggesting that the ED might listen to and act upon the efficiency opportunity if the information was relayed by someone he trusted. I hoped to use the concept of opinion leadership to convey the offerings in a more effective manner. *Opinion leaders* are generally thought of as individuals in a community whose opinion can influence the viewpoint of others. I thought using another nonprofit ED, who I had previously completed work with, as a reference, would be a good start to finding opinion leaders. Two of my contacts at separate nonprofits agreed to serve as references, which I provided to the reluctant ED. A few days later, I received a fax including agreement forms and a thank you. The installation has since been completed.

Every once in a while, I'll see the ED around town. Last time I saw him, he said he was still shocked about my program, and that if it hadn't been for the references he probably would have never moved the project forward! In this case, my observations, field notes, and understanding of sociological concepts and theories like opinion leaders and diffusion theory helped produce a more energy-efficient end. At the end of the day, the opinions of his peers held more weight in the determination of his decision than those of my organization, the state, or the local utility. The organization is now saving a lot of money and helping the world by reducing its carbon footprint.

In another example, my field notes, observations, and understanding of the barriers to CFL use, including risk perception, helped drive a lighting upgrade forward. I conducted an assessment for a business manager who had an energy-saving opportunity. He brought me in to take a look at his fluorescent lighting, but it turned out that he also had high-wattage incandescent lights. I recommended immediately changing the incandescent lights to CFLs. I told him RCEA was willing to provide the materials and labor for free because it could save so much energy. The manager told me he knew he could save money and energy but was afraid of the change because CFLs contain mercury (a small amount). He had heard on the news that it could cost a fortune to clean one up should it break. I knew of the exact headline he was referring to: A woman broke a CFL in her home, and it resulted in an expensive cleanup. In light of the episode, the United States Environmental Protection Agency (EPA) adopted new guidelines, which will reduce the likelihood of mercury exposure and an expensive cleanup should a CFL break. Regardless of the validity of his concerns, I knew and still know that the risk is worth taking because, in general, the use of the technology can reduce net environmental mercury output through energy savings while mitigating greenhouse gases like carbon dioxide.

As I did with the nonprofit ED, I referred back to my observation notes spanning several separate interactions. I identified the manager's perception of the risk associated with using CFLs as a potential barrier to the project moving forward. After reviewing my notes located in the database I had designed, I decided that I would go back to him and try to address his fear with information that clearly shows the CFL is an acceptable individual risk, considering the deeper societal consequences of not reducing electricity consumption.

A few days later, I returned to his business to present my findings. My hopes were that I could address the associated risks of electricity generation and CFLs. I explained that, in general, overall reductions in energy use associated with CFLs would save his business money; mitigate climate change; and reduce the number of rivers dammed, mountains leveled, and environmental tragedies like the 2008 Tennessee Valley Authority (TVA) coal ash disaster where coal ash sludge broke through a holding pond dike, flooding several hundred acres of land with toxic material. I also provided him with the EPA's CFL cleanup guidelines and showed him what was recommended. I finished by saying that CFLs are not a perfect answer—they do have inherent problems—but the societal benefits of reduced energy consumption outweigh the individual and communal risks of using the technology for now.

Later that day, the manager called and said he was willing to move forward with the change. I ordered the lights and installed them a week later. He has yet to break one and has told me he is less concerned about the possibility of doing so after seeing how much money and energy he's saving. The result is that the manager is now positively contributing to the reduction of greenhouse gases and the mitigation of climate change.

As the examples above depict, sociology has given me the tools to observe, note, synthesize, and make practical use of field observations and research. These tools include, but certainly are not limited to, a well-rounded understanding of diffusion theory and risk perception, a socio-logical imagination (realizing that one company's energy consumption is connected to the global issue of climate change), qualitative and quantitative research methodologies, and a good bit of critical thought. It is clear to me that without my sociological toolkit, I would be less effective in addressing energy efficiency and climate change. My education and use of sociology has helped me become an agent of real change. Sociology helps me to play a role in mitigating climate change and environmental degradation associated with electricity consumption every day!

ACTIVIST SCHOLARSHIP FOR ENVIRONMENTAL JUSTICE

David Naguib Pellow

University of Minnesota, Minneapolis

David Naguib Pellow is a professor of sociology at the University of Minnesota. His teaching and research focus on environmental justice issues in communities of color in the United States and globally. Among his books are *Resisting Global Toxics: Transnational Movements for Environmental Justice* (2007); *The Silicon Valley of Dreams: Environmental Injustice, Immigrant Workers, and the High-Tech Global Economy* (2002, with Lisa Sun-Hee Park); and *Garbage Wars: The Struggle for Environmental Justice in Chicago* (2004). He has served on the board of directors at the Center for Urban Transformation, Greenpeace, and International Rivers. He is the director of the Minnesota Global Justice Project.

Transatlantic Initiative on Environmental Justice

Environmental injustice occurs when any population suffers a disproportion-ately high burden of environmental harm and is excluded from environmental decisions affecting their community. The populations most affected include people of color, low-income communities, indigenous people, and women. A great deal of the sociological research on environmental injustice has emerged from collaborations between scholars and activists, and I have drawn inspiration from these models of political education and activist scholarship.

Most of the scholarship on environmental injustice focuses primarily on the United States and secondarily on communities in Africa, Asia, and Latin America. Until very recently, environmental racism in Europe was not given much attention. However, scholars, activists, and lawyers from environmental and human rights organizations in Bulgaria, Czech Republic, Hungary, Macedonia, Romania, and Slovakia have now come together to form the Coalition for Environmental Justice (CEJ) in Central and Eastern Europe (CEE). This coalition was developed with the expressed goal of combating environmental racism and human rights abuses directed at the Roma people of Europe. The Roma have, for centuries, been one of Europe's most despised and targeted ethnic groups, suffering continuous personal and mass violence at the hand of nation-states, institutions, and dominant cultural groups and individuals since the Roma migration into the area began in the 12th century. The Roma confront numerous environmental injustices in the region, including being forced to live on or near municipal waste dumps, poor access to clean water and sanitation, exposure to toxics (e.g., abandoned mines), and exposure to floods (e.g., living on islands or on rivers). As Larry Olomoofe, a Human Rights Trainer for the European Roma Rights Centre told me, "If you put your finger anywhere on a map of Europe where Roma are located, you'll find environmental problems."

CEJ members decided that they would benefit from an exchange with U.S. environmental justice scholars, activists, and legal advocates, with the goal of launching an effective global initiative to promote environmental justice in the region. This project was the first of its kind, in that it specifically linked the situation of vulnerable peoples and threatened environments in Central and Eastern Europe to the struggles of peoples in the United States. I was asked to coordinate the U.S. delegation to the exchange meeting.

In October 2005, I organized a group of academics, activists, and lawyers from the United States to travel to Budapest, Hungary, to join with counterparts from CEE nations. The workshop was held at the Central European University and focused on reporting on case studies, building strategies for policy and legal change, and creating networks among scholars and activists for future collaborations. The group eventually decided to call itself the Transatlantic Initiative on Environmental Justice (TIEJ), which would serve as a network linking advocates on both sides of the Atlantic, sharing critical information, developing new knowledge, and supporting campaigns.

The TIEJ then decided to move forward to support a number of campaigns and to initiate collaborative projects. One of the outcomes of the meeting was a human rights campaign to bring attention to a dire situation facing Roma populations in Kosovo who—after having been displaced by the Kosovo war in 1999—had been relocated by the United Nations (UN) to a toxic waste site where children were now suffering from lead poisoning (Wood, 2006). More than 500 Roma were moved to northern Kosovo to escape the war. Unfortunately, they were placed in refugee camps where they were exposed to severe lead poisoning because the sites were located near a major mine and lead smelter. The World Health Organization (WHO) conducted a study in these camps and found that 88% of the children under age six had blood lead levels in the highest category, described as an "acute medical emergency" (Global Response, 2005).

For many years, I have worked with Global Response—an organization that mobilizes letter-writing campaigns to bring attention to human rights and environmental justice abuses around the world. This letter-writing work focuses public attention on leaders, governments, and corporations involved in policy making associated with environmental and social harms, but it is also an important venue for creating and disseminating alternative knowledge that seeks to redefine a situation as an injustice. Drawing on this resource, I brought together Global Response (Colorado) and the European Roma Rights Center (Budapest) to launch an international letter-writing campaign to persuade the United Nations to relocate Roma families from the toxic waste site in Kosovo. After hundreds of letter writers from around the world participated in this effort, the campaign secured an unprecedented response from the UN, which moved to create a relocation plan within a month's time. I also helped to create an electronic Web site that allows activists involved in this network to electronically publish their case studies of community struggles for environmental justice, and we have an Internet listserv that facilitates regular communication and information sharing across national borders. These are just two of the many projects that emerged from this exchange.

The TIEJ speaks directly to the power of political education and collaborative action by university scholars and community leaders to transform both the academy and the world around us. It is also a collaboration focused primarily on the power of sociological ideas to facilitate the production of knowledge for social change. Specifically, this case reveals the fact that environmental crises are always socially constructed and deeply reflective of social inequalities. That is, the environmental *crises* that Roma people face are, in many ways, environmental *solutions* for the non-Roma majority of Central and Eastern Europe. The dominant group's pollution is simply placed in Roma communities so that it is out of sight and out of mind. This situation only becomes a crisis for the dominant group when marginal populations like the Roma disrupt the social relations that produce environmental inequality and make those practices a general social problem. Drawing on the sociology of knowledge and linking it with collective action, the TIEJ was able to make visible the otherwise hidden crisis of environmental racism in Roma communities.

Applying Sociological Approaches to Mining Conflicts

Building on the TIEJ project, in 2009 I again joined forces with Global Response to produce a guide for communities facing threats from mining companies engaging in predatory practices. For years, Global Response has defended communities whose political rights and ecosystems are under fire from mining companies that seek to siphon off natural resources for profit, while leaving local people with little in return. Large mining operations have polluted waterways, air, and land in affected communities around the globe (Gedicks, 1993). Deforestation, fish kills, and the reduction of wildlife habitat and biodiversity frequently accompany the construction and operation of large mines (Gedicks, 2001). The social impacts of such facilities can include human rights violations (intimidation, torture, rape, and murder) directed at activists who dare to oppose mines, the loss of livelihood associated with the destruction of farmland and fisheries, respiratory illnesses, cancer, birth defects, radioactive exposure, black lung disease, malnutrition, and the loss of cultural resources (such as sacred sites for religious ceremonies, including mountains, bodies of water, and trees) (LaDuke, 2005). In short, while our modern world has grown dependent on mining to produce our oil, coal, metals, and minerals, the social and ecological toll of these practices is quite high indeed. And since the people on the front lines of these resource wars need all the help they can get, I was pleased to join in the fight as an activist-scholar who could apply concepts from social movement theory, such as collective action frames, resource mobilization, and political opportunity structures to a real-world situation. These concepts from social movement theory explain how communities of resistance articulate their concerns and how they mobilize most effectively against their targets in order to create a better world for themselves.

So, Global Response and I teamed up with DECOIN, an environmental justice group in Ecuador that has faced down some of the world's most powerful mining companies in that country, to author a guide on how to strengthen community struggles against mining corporations (Zorilla, Buck, Palmer, & Pellow, 2009). The report is intended for community organizers and anyone else interested in learning about how ordinary people can confront large mining companies that attempt to take control of communities to ensure profits for their shareholders, no matter the cost to the local people and ecosystems. We present several cases that detail how activists can fight back and win, by doing the following: building alliances and coalitions, using the law and the political process, framing messages and using the media, organizing globally, and engaging in direct action. The guide includes a list of common mining company tactics to divide communities and to pressure governments to grant them access to natural resources, along with actual examples of how activists have countered them. Finally, we also provide a list of resources that contains the names of nongovernmental organizations around the world that are working on mining justice campaigns, technical and legal information, and a special section on how grassroots campaigners can use the media to their advantage.

The guide's impact has been immediate. Since it was released in 2009, activists in Europe, Canada, East and West Africa, Latin America, Asia, and the United States have used it in their campaigns. An activist from Sierra Leone told me, "My community is fighting a mine right now and this guide is a great help to us." A U.S.-based activist working on mining conflicts declared, "This is the best guide on defending communities against mines I have ever seen."

I now routinely apply sociological concepts from environmental justice studies, the sociology of knowledge, and social movement theory to community struggles around the world. I plan on recruiting sociology students to assist me with research for the next edition of the mining guide and on a research project focused on Roma communities working toward environmental justice in Europe. Each of these experiences reminds me of the power of sociology as a field where ideas and action can come together to produce new knowledge and social change. When people ask me why I became a professor of sociology, I always say, "Because I wanted to change the world. Why else?"

References

Gedicks, A. (1993). *The new resource wars: Native and environmental struggles against multinational corporations*. Boston, MA: South End Press.

Gedicks, A. (2001). *Resource rebels: Native challenges to mining and oil corporations*. Boston, MA: South End Press.

Global Response. (2005). *GR Action #4/05*. Retrieved from http://www.globalresponse.org

LaDuke, W. (2005). *Recovering the sacred: The power of naming and claiming*. Boston, MA: South End Press.

Wood, N. (2006, February 5). Displaced gypsies at risk from lead in Kosovo camps. *New York Times*. Retrieved from http://query.nytimes.com/gst/fullpage.html?res=9F03E7DE173EF936A35751C0A9609C8B63&sec=&spon=&pagewanted=2

Zorrilla, C., Buck, A., Palmer, P., & Pellow, D. (2009). *Protecting your community against mining companies and other extractive industries*. Boulder, CO: Global Response and the Minnesota Global Justice Project.

THE SOCIOLOGY OF ENVIRONMENTAL JUSTICE: MERGING RESEARCH AND ACTION

Daniel Faber

Northeastern University, Boston, Massachusetts

In 2006, Daniel Faber was recognized as a "Champion of the Earth" by Salem State College and HealthLink, as well as "Outstanding Environmental Advocate of the Year" by the Alliance for a Healthy Tomorrow, a statewide environmental coalition of over 160 organizations in Massachusetts. Dr. Faber has also been awarded a Certificate of Appreciation by the Environmental Protection Agency, as well as by the National Association for the Advancement of Colored People (NAACP) at Northeastern University, for advancing environmental justice in Massachusetts and beyond. His book, *Capitalizing on Environmental Injustice: The Polluter-Industrial Complex in the Age of Globalization* (2008), was a finalist for the 2009 C. W. Mills Award.

As an activist-scholar, my life is devoted to advancing a sociological analysis of the world's most pressing environmental issues. I believe that such an approach is necessary because environmental activism can be effective only when guided by sociological prescriptions that take into account the interconnections between social and ecological issues. As stated by Pablo Eisenberg (1997) of Georgetown University's Public Policy Institute, "Although we know that our socioeconomic, ecological, and political problems are interrelated, a growing portion of our nonprofit world nevertheless continues to operate in a way that fails to reflect this complexity and connectedness" (p. 331). As a result, the linkages among environmental abuses, poverty and economic inequality, racism, human health problems, and political-economic power are typically ignored. For this reason, I seek to develop and integrate an environmental justice (EJ) perspective into my sociological work. The notion that "not all people are polluted equal" is a central concern in my ongoing research. My quest is to uncover the root causes of environmental inequities in the United States and around the globe. Furthermore, my work aims to provide policy makers, environmental advocates, scholars and scientists, social justice activists, foundation officials, and ordinary citizens with real solutions to the ecological crisis.

Sociological Research on Environmental Justice

Not all people enjoy the same protection from ecological hazards. Rather, industries and government agencies regularly adopt pollution strategies that offer the path of least resistance: displacing ecological harm onto the public in ways that are politically "expedient." As a result, the less political-economic power a community possesses to defend itself, the more likely it is to suffer arduous environmental and human health problems. Throughout the United States, it is communities of color, industrial laborers, rural farm workers, immigrant labor, and working-class neighborhoods that are being harmed to the greatest extent. A report by Cerrell Associates (1984) for the California Waste Management Board, for instance, recommends that "middle and higher-socioeconomic strata neighborhoods should not fall at least within the one-mile radii" of any proposed incinerator site (p. 42). Instead, the report recommends that the state target "lower socioeconomic neighborhoods"—which they defined as primarily low-income, rural, or Catholic—because those communities had a much lower likelihood of offering opposition. In Greater Los Angeles, 91% of the 1.2 million people living in close proximity (less than 2 miles) to the city's 17 hazardous waste treatment, storage, and disposal facilities (TSDFs) are people of color.[3]

My own research is concerned with the disparate exposures to environmental health hazards experienced by people of color and working-class whites. This first takes the form of higher rates of on-the-job exposure to toxins used in the production process, and second as greater neighborhood exposure to pollutants emitted from nearby factories, toxic dumps, agricultural fields, transportation systems, and hazardous waste facilities. Third, unequal exposure to ecological hazards manifests as unequal enforcement or faulty cleanup efforts implemented by the government or the waste treatment industry, such as through the increased use of permanent or mobile incinerators that burn this waste in the community. For instance, government penalties for violations of Superfund hazardous waste laws in communities of color average only one-sixth ($55,318) of what they do in predominantly white communities ($335,566).[4] The final piece to the quadruple exposure effect comes in the form of greater exposure to toxic

---

[3]Statewide, 81% of the people living in close proximity to TSDFs are racial/ethnic minorities. Robert D. Bullard, Paul Mohai, Robin Saha, and Beverly Wright, *Toxic Wastes and Race at Twenty: 1987–2007—Grassroots Struggles to Dismantle Environmental Racism in the United States*, a report prepared for the United Church of Christ Justice and Witness Ministries (March 2007), 58–60.

[4]See Marianne Lavelle and Marcia Coyle, "Unequal Protection: The Racial Divide in Environmental Law" (*National Law Journal*, September 21, 1992), 2–12.

chemicals in the household (such as lead paint), commercial foods, and a variety of consumer products. For example, lead poisoning continues to be a leading health threat to children, particularly poor children and children of color living in older, dilapidated housing. Black children are now five times more likely than white children to have lead poisoning. Taken together, it is clear that people of color and working-class families experience a disparate exposure to environmental hazards where they "work, live, and play" (Bullard, 2005). This is true internationally as well (Pellow, 2007).

In 2001, Dr. Eric Krieg and I released a report on ecological stratification in Massachusetts (we updated and expanded the report in 2005). Entitled *Unequal Exposure to Ecological Hazards: Environmental Injustices in the Commonwealth of Massachusetts,* the study is unique in a number of ways. Most other EJ studies focus on limited geographical areas (such as a city), or if national in scope, analyze only one type of hazardous facility or waste site. In such circumstances, it is difficult to claim disparate exposure based upon studies of only one type of facility, or in a very limited range of communities of different sizes and populations. Most of these studies also fail to develop a method for measuring the relative severity of the ecological hazards faced by different communities vis-à-vis each other. As a result, we are left with a limited picture of the total environmental burden faced by people of color and working-class families. Our study is the first to bridge these shortcomings, and is, to date, the most comprehensive environmental justice study of any state in the country.

Our report analyzes both income-based and racially based biases in the geographic distribution of some 17 different types of environmentally hazardous sites and industrial facilities in every single community in Massachusetts. We also measure the density of ecological hazards within and between communities (i.e., frequency and severity of hazards per square mile), which allows us to consider the different sizes and populations of these communities. For instance, our study reveals that communities of color in the state average an eye-popping 48.3 hazardous waste sites per square mile. In contrast, white communities average only 2.1 hazardous waste sites per square mile. As a result, *communities of color average 23 times more hazardous waste sites per square mile than white communities.* We similarly analyze the total industrial emissions of carcinogens, persistent bioaccumulative toxins (PBTs), and reproductive toxins in every community in Massachusetts between 1990 and 2002. In so doing, we uncover that communities of color receive well over 33% of all the carcinogens, PBTs, and reproductive toxins released by industry (but comprise only 9.4% of all communities).

Moreover, we developed a system for ranking the cumulative environmental burden in each community. Our findings indicate that ecologically hazardous sites and facilities, ranging from highly polluting power plants to incinerators to toxic waste dumps, are disproportionately located in communities of color and working-class communities. In fact, the numbers are very disturbing: Low-income communities face a cumulative exposure to environmentally hazardous facilities and sites that is *four times greater* than high-income communities. Far worse, communities of color experience a *cumulative* exposure rate to environmentally hazardous facilities and sites that is *over 20 times greater* than white communities. We conclude that striking inequities in the distribution of these sites and facilities place lower-income families and people of color at a substantially greater health risk. In fact, a person living in a community of color is *39 times* more likely to live in one of the 30 most environmentally hazardous towns in Massachusetts.

Action Based Upon Our Sociological Research on Environmental Justice

Prior to our 2001 study, the assumption among many policy makers was that environmental injustice was insignificant in Massachusetts and a problem confined to the Sunbelt and other parts of the country. Our report exploded that myth and generated enormous statewide and national media attention, including interviews with National Public Radio, CBS News, *Scientific American,* the *Boston Globe* and *Herald* newspapers, local television and radio stations, and countless other media outlets. More importantly, our report fueled organizing efforts by activists and served to jump-start a lagging legislative process to adopt a draft EJ policy for Massachusetts. As stated by Veronica Eady (2003) of the Executive Office of Environmental Affairs (EOEA), and a board member of the National Environmental Justice Advisory Council (NEJAC) at the Environmental Protection Agency,

> The Faber/Krieg report . . . sent shockwaves through neighborhoods across Massachusetts where residents sensed an unfair pollution burden but did not know what to call it. All of a sudden, "environmental justice" was a widespread battle cry, not just across the state's communities of color and low-income neighborhoods, but all across the landscape. The Faber/Krieg report created a media splash that produced two key results. First, residents began to look closely at the draft environmental policy. . . . The Faber/Krieg report drew careful scrutiny to the draft policy not only by residents, not only by the media, but also by other states and by industry. The second key development induced by the Faber/Krieg report was that the activists demanded legislation that would in effect make an environmental justice policy adopted by the EOEA applicable to state agencies across the board, not just environmental agencies. (p. 171)

Media furor surrounding our report led to intense pressure on the EOEA to adopt an EJ policy, and led me to testify before the Joint Committee on

Natural Resources and Agriculture in the State House on May 17, 2001. A few months later, the legislature and governor supported implementation of an EJ policy. The new EJ policy included many of the recommendations made in our report, and is now among the most comprehensive in the nation. In fact, Massachusetts Governor Deval Patrick, who extensively cited our report in his own environmental platform during his 2005–2006 campaign for governor, is currently considering an executive order on environmental justice that would strengthen the gains already made. The executive order is being pushed for by the Massachusetts Environmental Justice Alliance, of which I am a cofounding member.

A final word: My sociological investigations tell me that organizing efforts against the procedures that result in the unequal *distribution* of environmental problems have limited success. Attempts to rectify distributional inequities without attacking the fundamental processes that produce the problems in the first place are largely focused on symptoms rather than causes and are only a partial, temporary, and necessarily incomplete solution to environmental injustice. We need additional policies that emphasize greater public participation in the capital investment decisions through which environmental burdens are *produced* and then distributed. This includes state programs and policies that mandate the phase-out of dangerous chemicals and the phase-in of safer substitutes, cleaner technologies and production processes, and a more precautionary approach to environmental problem solving.

In Massachusetts, I helped to cofound the Alliance for a Healthy Tomorrow (AHT), a coalition of environmentalists, public health advocates, labor unions, faith-based organizations, scientists, EJ activists, nurses and doctors, and ordinary citizens working for this more production-focused approach. The AHT has successfully won a phase-out of mercury in Massachusetts, and is advocating for a number of pieces of legislation. Among these is an Act for a Healthy Massachusetts, which would mandate the phase-out of dangerous chemicals for use by industry. Instead, businesses would be required to adopt safer substitutes for these dangerous chemicals. If adopted, we would be the first state in the country to have this type of legislation. I have also utilized my sociological training to serve as one of the architects of the Safer Alternatives bill, which I am hopeful will come up for a vote within the next year before the full legislature. But more needs to be done. As director of the Northeastern Environmental Justice Research Collaborative, I am working on new sociological research and action initiatives around climate change, globalization, philanthropy, and other environmental justice issues. The urgency of the global ecological crisis demands it of me, as it does of all sociologists.

References

Bullard, R. (Ed.). (2005). *The quest for environmental justice: Human rights and the politics of pollution.* San Francisco, CA: Sierra Club Books.

Cerrell Associates. (1984). *Political difficulties facing waste-to-energy conversion plant siting.* Report for the California Waste Management Board. Los Angeles: Author.

Eady, V. (2003). Environmental policy in state policy decisions. In J. Agyeman, R. Bullard, & B. Evans (Eds.), *Just sustainabilities: Development in an unequal world.* London, UK: Earthscan Publications.

Eisenberg, P. (1997, Winter). A crisis in the nonprofit sector. *National Civic Review, 86*(4), 331–341.

Faber, D. R., & Krieg, E. J. (2005). *Unequal exposure to ecological hazards 2005: Environmental injustices in the Commonwealth of Massachusetts.* A Report by the Philanthropy and Environmental Justice Research Project, Northeastern University, Boston. Retrieved from http://www.socant.neu.edu/research/justice_research/publications/

Pellow, D. N. (2007). *Resisting global toxics: Transnational movements for environmental justice.* Cambridge: MIT Press.

DISCUSSION QUESTIONS

1. Andrea Rother describes how her experience in southern Kenya showed how the team with which she was working "lacked the essence of sociological research." What did she mean by that and what was she, as a sociologist, able to bring to the project that these other researchers lacked? How did her sociological tools bring about the "salvation of the project"?

2. Rother's work to curb the selling of street pesticides in and around Cape Town, South Africa, provides a powerful example of how sociologists can make a positive impact on society. How was she able to use her sociological tools to provide evidence of the problem and effectively advocate for appropriate intervention strategies? If you could hire Andrea Rother to conduct research in your local area, what would you ask her to do? Why?

3. Do you make it a point to conserve energy? Why or why not? Has Jacobson's Sociologist in Action piece influenced your thinking? If so, how might you use sociological tools to convince someone to actively fight environmental pollution and global warming? If not, why not, and what might it take for you to consider changing your energy use habits?

4. Jacobson says that his drive to combat global warming and environmental degradation stems from growing up in the heart of coal country in the

Appalachian Mountains. How is this an example of his use of the sociological imagination? Can you think of a personal trouble of your own that might spur you to work for energy conservation? How do you think (a) where you live, (b) your age, and (c) your social class impacted your answer?

5. Pellow maintains that "environmental injustices are socially created." What does he mean by this statement? What are some examples that help to prove his point?

6. Pellow describes how he works for environmental justice with colleagues all over the world. Why do you think it is essential for us to look at environmental justice as a global issue? How does the fact that we now have a global work-force (with corporations moving across national lines in search of cheap labor) play a role in the struggle for environmental justice?

7. Why did Faber's work result in "shockwaves" across Massachusetts? Were you surprised to find that environmental injustice exists in a state like Massachusetts? Why or why not? If you were a citizen from Massachusetts (or if you are), how might you use Faber's findings to work toward reducing environmental injustices?

8. What does Faber mean when he says we need a "more precautionary approach to environmental problem solving"? How might we go about achieving such an approach? How might you and fellow students take this type of approach on your campus or in your community?

9. Is there a student organization on your campus that works for environmental justice? If not, why do you think one doesn't exist? What might lead to one being created? If there is one, do you think it is having a positive impact on society? Why or why not? If yes, what makes it effective? If no, what recommendations might you make to help it become more effective? How might you, personally, contribute to such an effort?

10. Thirty years ago, few, if any, sociology textbooks contained chapters on environmental justice. Today, almost every sociologist recognizes the importance of addressing (and teaching about) the connection between social and ecological issues. What do you think brought about this change?

RESOURCES

The following Web sites will help you to further explore the topics discussed in this chapter:

ASA Section on Environment & Technology	http://www.envirosoc.org/
Environmental Justice/Environmental Racism	http://www.ejnet.org/ej/

EPA Climate Change	http://www.epa.gov/climatechange/
EPA Environmental Justice	http://www.epa.gov/environmentaljustice/
Global Warming: Early Warning Signs	http://www.climatehotmap.org/
New York Times Global Warming Page	http://learning.blogs.nytimes.com/teaching-topics/global-warming/#
Sociosite Environment—Ecology	http://www.sociosite.net/topics/environment.php
Story of Stuff	http://www.storyofstuff.com/

To find more resources on the topics covered in this chapter, please go to the Sociologists in Action Web site at **www.sagepub.com/korgensia2e.**

Chapter 13

Social Institutions (Family, Economy)

Social institutions are patterns of behavior governed by rules that are maintained through repetition, tradition, and legal support. Members of every society create institutions to control human behavior and go about meeting their basic human needs. The five major social institutions are the *family*, the *economy*, *education*, *government*, and *religion*. In this chapter, we look at the work of Sociologists in Action who study the social institutions of the family and the economy.

In "Sociological Advocacy for Children," Yvonne Vissing describes how she brings sociological insight into her work for children. Her sociological training enables her to realize that changes in one major social institution impact the other institutions. Seeing families as the "foundation" of society, she has worked to strengthen the institution of the family and improve the lives of children through creating the Center for Child Studies at Salem State University. All of Vissing's efforts stem from her vision of a better world for children. Through her research, advocacy for homeless children and at-risk youth, and her work training undergraduates to become child advocates, she has made this vision a reality for many children.

Patrick Carr and Maria Kefalas share findings from their research on small-town Middle America in their piece, "From Hollowing Out the Middle to Reimagining Small Towns." Using data collected through interviews and participant observation, Carr and Kefalas carefully and vividly document the pattern of high-achieving young residents moving away from small-town America and the repercussions of this pattern on the economic viability of their hometowns. Carr and Kefalas note that while "this so-called 'brain drain' is not a new phenomenon, its effects are especially debilitating in a modern economy that has mechanized agriculture and outsourced most manufacturing." They believe that to combat this problem, small towns must be "reinvented," and they have used their findings in the public media and in other outlets to suggest how such a "reinvention" may take place.

Leslie Hossfeld concludes this chapter with "'Why Don't We Do Something About It?' A Response to Job Loss in Rural Communities." When giving a lecture one day to students in rural North Carolina about the impact of the movement of factories from the United States to Mexico and other places with cheaper labor, it suddenly struck her that her students were living what she was speaking about! That experience spurred Hossfeld to work in collaboration with others in the community and her students to document the decline in jobs in the area, report their findings to Congress, and advocate for specific policies to address the situation. Those actions led to the creation of a national conference on job loss and a variety of projects aimed at "economic recovery for the region and rural America, in general."

SOCIOLOGICAL ADVOCACY FOR CHILDREN

Yvonne Vissing

Salem State University

Yvonne Vissing, PhD, is the founder and director of Salem State University's Center for Child Studies and Professor of Sociology. Author of a half dozen books and dozens of articles, chapters, and professional presentations in the United States and internationally, she has dedicated her life to promoting child well-being. She is a board member of the National Coalition for the Homeless, gubernatorial appointee to the New Hampshire Juvenile Parole Board, and a National Institute of Mental Health Post-Doctoral Research Fellow. She was recently awarded Sociologist of the Year by the New England Sociological Association.

"Why are you majoring in sociology? What kind of job can you get with that degree?" These questions were often asked of me when I was a student. I had no idea when I majored in sociology how it would change my life and enable me to change a little piece of the world. I rely upon sociological skills to work with organizations in hopes of making the world a better place. These skills help me to have better interactions with others. They enable me to analyze situations and create plans for change. They also help me to understand myself. I use them almost every hour of every day. For me, sociology is not just an academic discipline but a framework for life. For most of those whom sociology calls, we enter the field not quite confident how we will use it, but sure that it has an enlightened vision that will somehow make life more worthwhile.

I did not understand, as an undergraduate, that the journey to a good career and life destination is seldom clear and easy. Like athletes in training, we can't accomplish our professional and personal goals unless we have the strength and skills to do so. It can't happen overnight; it takes time to build muscles in the body—and in the mind. Who we are and who we will become are shaped by time, place, and mind space. The impact of demographics and culture is also inescapable.

I grew up in Indiana during the 1960s, when my small town was transitioning from a farming community into a bedroom suburb of a metropolitan area. Longstanding rural values mixed with the values and lure of cosmopolitan possibility. My middle-class parents never went to college and the accepted norms of the day were based on traditional, circa 1950, gender roles. My father thought if I went to college at all, I should become a secretary. My mother, however, pushed me to go to "real college" which she had longed to attend. My parents couldn't mentor me through my academic journey because they didn't understand the world of higher education, but they did socialize me to understand that everyone should be treated equally and they taught me the importance of fighting for the underdog. This turned out to become a lifelong theme in my work.

The world was exploding with social change when I was in college and I majored in sociology because it gave me a way of understanding what was happening. I also thought the degree would enable me to help others. A family friend helped me to secure a job as a child abuse prevention specialist in a mental health center. It was an interesting job and I did help others, but it did not pay well. Organizational cost-cutting soon eliminated prevention jobs. I decided to go to graduate school where I earned my PhD in sociology.

Upon completing graduate school, I took a teaching position in a small Midwestern college, had children, bought a house with a picket fence and got a dog. It was a good life but I wanted to work in a more supportive research environment. I became a visiting professor at a major Midwestern university with a joint appointment in sociology and the medical school. Then I became a Postdoctoral Research Fellow and moved to the Boston, Massachusetts, area where I began to commit myself seriously to work for the underdog, in the form of children and people who are homeless.

Sociology assumes that an open system exists between institutions; changes in one naturally affect the others. Vulnerable individuals are dependent upon social institutions to adequately care for them. For example, children depend upon how their parents care for them; children think parents have power, but parents' lives are influenced by a host of forces over which they have little control. The time and money parents have to spend on children is impacted by their jobs. Meanwhile, their employers are reacting to larger economic forces which are influenced by decisions that political leaders make, who claim to reflect the will of the people. A variety of institutions exist to help children and youth—but how well do they do it? Often, not well enough because the social structure fails to adequately invest in children.

In response to the need for more child-centered infrastructure, I created the Center for Child Studies at Salem State University (SSU) in 1999 to teach students about the importance of using their academic skills to improve the lives of children. We have since trained thousands of students to become child advocates and work with a variety of community groups and organizations on behalf of children.

I believe that if we choose to build the human infrastructure by investing today in children, we could change the world for the better in a single generation! Imagine what would happen if an entire generation of children and their parents were given good health care, quality nutrition, adequate and affordable housing, secure family incomes, supportive parenting programs, quality education, violence-free communities, positive recreational outlets, and positive social opportunities. These children would grow into adults well-equipped to meet the social, political, and economic needs of the future. They, in turn, would be better equipped to take care of their children, who would then be better caretakers of the world. This is how one builds a strong society.

Failing to address issues of injustice, discrimination, deprivation, alienation, and isolation are sure-fire ways to ensure that we create the kind of world that isn't good for anyone. Society is built by investing in the human infrastructure, not just the buildings, roads, or the economic infrastructure. Families are the foundation of any society's infrastructure.

My childhood upbringing instilled in me the belief that we are our brother's keeper and that I had an obligation to make the world a better place. At first, though, I didn't know how to do that. Sociology provided me with a roadmap. The professional socialization I received transformed me into a pediatric, applied sociologist who is improving the lives of children. The academic core provided me research skills, theoretical insights, and knowledge of social problems and deviant behavior. I learned about how organizations operate and why people think and act the way they do.

While having a good heart is important, I learned that one also needs a good head in order to create positive social change.

I love research and have conducted many projects on homeless children, which led to my first book, *Out of Sight, Out of Mind: Homeless Children and Families in Small Town America*, and the award winning video, *I Want to Go Home*. It is amazing how the public's knowledge about poverty and homelessness is so far askew from the facts. As a board member of the National Coalition for the Homeless (NCH), I try to use data to educate the public and policy makers. I have had students transform NCH archives into abstracts that anyone can use by accessing the NCH and SSU library Web sites (http://www.noblenet.org/ssu/research/internet_resources/NCH/). This database is being used by scholars and policymakers to create research and presentations that promote social change.

As a member of the New Hampshire Juvenile Parole Board, I meet with troubled youth, their families, law enforcement, and social service providers to craft better lives for youth-at-risk. I work with the courts as a Guardian ad Litem in child custody and divorce cases to ensure that children's needs get addressed. I consult with schools, medical facilities, social service organizations, and the legal community to promote child and family well-being.

I use my sociological imagination to envision a world in which all children have their needs met. It is my belief that we have the opportunity and obligation to help children, but this can only occur when children become a priority. This is why we as sociologists have to advocate for them. Sociology has given me a theoretical framework and pragmatic skills to help create the kind of wonderful world that I want for my children, and all the children of the world.

References

Vissing, Y. M. (2011). *Introduction to sociology*. San Diego, CA: Bridgepoint Education Publishing.

Vissing, Y. M. (2007). *How to keep your children safe: A guide for parents*. Lebanon, NH: University of New England Press.

Vissing, Y. M. (2002). *Women without children: Nurturing lives*. New Brunswick, NJ: Rutgers University Press.

Vissing, Y. M., & Peer, S. (2001*). Finding information about children: Using human and electronic resources*. Hauppauge, NY: Nova Science.

Vissing, Y. M. (1996). *Out of sight, out of mind: Homeless children in small town America*. Lexington: University of Kentucky Press.

Vissing, Y. M. (2010). Curfews. In W. Chambliss (Ed.), *Juvenile crime and justice: Key issues in crime and punishment*. Thousand Oaks, CA: Sage.

Vissing, Y. M. (2007). Rhetoric of concern: Child poverty and homelessness in the USA. In R. Findlay & S. Salbayre (Eds.), *Stories for children, histories of childhood/ Histoires l'enfant, histoires d'enfance*. Tours, FR: GRAAT (Groupe de Recherches Anglo-Américaines de Tours, EA 2113).

Vissing, Y. M. (2007). A roof over their head: Applied research issues and dilemmas in the investigation of homeless children and youth. In A. Best (Ed.), *Representing youth: Methodological issues in critical youth studies*. Thousand Oaks, CA: Sage.

Vissing, Y. M., Straus, M., Gelles, R., & Harrup, J. (1991). Verbal aggression by parents and psychosocial problems of children. *Journal of Child Abuse and Neglect 15*(3). Retrieved from http://pubpages.unh.edu/~mas2/VB35C1.pdf

Vissing, Y. M. (Winter 2003). The Yellow School Bus Project: How religious and community organizations can prepare homeless children for school. *Phi Delta Kappan*. Retrieved from http://www.kappanmagazine.org/content/85/4/321 .abstract

Vissing, Y. M. (Spring 2003). The subtle war against children. *Fellowship*. Retrieved from http://www.forusa.org/fellowship/marapr_03/vissing.html

Vissing, Y. M., & Hudson, C. (2008, November 2). Issues in enumerating homeless children and youth. Paper presented at the National Association for the Education of Homeless Children and Youth Conference, Crystal City, VA. Abstract retrieved from www.naehcy.org/conf/dl/2008/vissing_issues.doc

FROM HOLLOWING OUT THE MIDDLE TO REIMAGINING SMALL TOWNS

Patrick J. Carr

Rutgers University, New Brunswick, New Jersey

Maria J. Kefalas

Saint Joseph's University, Philadelphia, Pennsylvania

Husband-and-wife sociologists Patrick J. Carr (Rutgers University at New Brunswick) and Maria J. Kefalas (Saint Joseph's University, Philadelphia) are authors of *Hollowing Out the Middle: The Rural Brain Drain and What It Means for America* (2009). Both have authored numerous other works, including Patrick J. Carr's *Clean Streets: Controlling Crime, Maintaining Order, and Building Community Activism* (2005) and Maria J. Kefalas's *Promises I Can Keep: Why Poor Women Put Motherhood Before Marriage* (2007, coauthored with Kathryn Edin).

Any scholar who is being totally honest will tell you that there is a certain amount of serendipity involved in every research project. Studies rarely unfold the way they are meant to, and, though this can lead to a great degree of anxiety while you are actually doing the research, it can also open up possibilities that can take your work in new and exciting directions. Certainly this has been our experience working on what we call the "Heartland Project," a study of young adults who grew up in a small Iowa town in the late 1980s and early 1990s (Carr & Kefalas, 2009). When we began the Heartland Study, we focused on several cohorts of high-school students in the Iowa town we call Ellis that had graduated in the late 1980s and early 1990s, the idea being that when we caught up with these young people, they would be in their middle and late twenties, and qualified to speak about what it is like to be an adult. To get a sense of the population of young adults, we distributed a survey to as many of the approximately 320 people who had attended Ellis High School during the targeted time frame as possible, and the responses we received (over 80% in all) came from young adults still living in the town itself (about 25%), elsewhere in the state of Iowa (28%), and all over the continental United States and beyond (47%). Fortuitously, we decided that we needed to interview the people who had stayed in Ellis and those who had settled elsewhere, and we quickly set about tracking people down to arrange in-depth conversations.

It was in these initial interviews that we saw our first signpost to what would be the final destination of our work. One of the first things we asked about in our interviews was where the young adults currently lived, where they had grown up, and what had prompted any major moves in their lives. It became readily apparent that anyone growing up in this small town faced two key questions when they came of age. First, do you stay or do you leave? Second, if you leave, do you ever come back? Those who left talked about the difficulty of doing so, but how they felt that no matter what they missed about the town they grew up in, there was just something drawing them away from the place. On the other hand, those who stayed said they really couldn't imagine living anywhere else. This latter sentiment was echoed by many who returned to Ellis, and they spoke about how they found the world beyond their town alienating, or how their time away had failed to live up to their expectations.

If the early interviews had hinted at a wider issue, our time actually living in Ellis in 2002 offered further clues as to what would become the focus of our writing. We moved to Ellis in the summer of 2002 to spend time in the town and to interview people who stayed or returned to live in the town. Summer also afforded us the opportunity to catch up with young people visiting family, especially during the three-day Ellis Summer Festival held each July. The experience living in the town helped us appreciate not just the local norms and mores of a small town, but also the peculiar challenges that such places face in the early 21st century. A few weeks into our stay there, the local Rotary Club called us up and asked if we wouldn't mind speaking to their weekly lunch gathering about our research. They, along with almost everyone else in Ellis, were curious about what we were doing, but, more importantly, they had a pressing issue that they wanted us to address. They explained that they wanted to figure out how best to attract young professionals, such as doctors, lawyers, or entrepreneurs to Ellis, and, as we had spoken to several local success stories that had left the town, they wondered if we had any insights as to what they should do. We duly gave our presentation on the Heartland Study, and in the conversation that followed, we tried to address the need to replenish the ranks of local professionals. Though it was early in our research, we had spoken to enough young adults who had left Ellis to pursue professional careers elsewhere to know that it would not be easy to get them to come back. In fact, if anything, most of them had told us that they had been carefully prepared to leave Ellis, being "set forth to do great things" as one young woman put it to us.

"Leavers" of Ellis were of two main types: the *achievers,* who were the ones carefully cultivated to succeed and spread their wings beyond Ellis, and *seekers* who had a strong desire to see the world but not the preparation of the achievers, and so joined the military as a ticket out of Ellis. The "stayers," by comparison, never left Ellis and worked in the remnants of the dying agro-industrial economy, feeling prepared for little else. As we charted the pathways that Ellis youth take as they wind their way to adulthood, we were struck by the key role that towns themselves play in this sifting and sorting mechanism that leads some to leave and others to stay in small towns.

Upon closer inspection, it became clear that the choices that many young people make are not accidental, but rather the result of years of careful mentoring and cultivation from parents, teachers, and community members. In the years since we lived in Ellis, we have gone back there several times and have come to the realization that there is a certain universality in the experiences of the young adults who came of age there. Towns like Ellis allocate their collective resources unevenly across groups of young people and tend to expend most of their efforts on cultivating those most likely to leave and not come back, while underinvesting in the people who are the future of towns. This so-called "brain drain" is not a new phenomenon, but its effects are especially debilitating in a modern economy that has mechanized agriculture and outsourced most manufacturing. For example, many of the stayers spoke poignantly of how they would be the first generation who would not be economically better off than their parents. The fact that these wider economic forces are aided and abetted by what happens at a local level makes this a difficult problem to combat, but one that is certainly not intractable.

The scale of what we came to call the "hollowing-out" problem is borne out by some wider trends. Between 1980 and 2000, over 700 nonmetropolitan counties in the United States lost 10% or more of their population, and further, between 2000 and 2005, there were more deaths than births in 1 out of every 2 rural counties (Carr & Kefalas, 2009). The population being lost in these places is predominantly educated youth, leaving an increasingly aging population and towns that have more difficulty attracting young professionals. The sheer scale of the problem, which cuts a swath through the American Heartland and affects places from Maine to Louisiana, is far greater than boom-and-bust trends of the 19th century, and will require a sustained effort to overcome. Luckily, with commitment and a plan, this problem is solvable if we can move to reimagine small towns.

To accomplish this reimagining, we have crafted a set of policy recommendations that focuses on changes at local, state, and national levels to help rebuild small towns (Carr & Kefalas, 2009). We exhort local leaders to reimagine education in such a way as to equalize resources across different groups of young people, basically advocating that they pay as much attention to stayers as achievers, and utilizing the existing community college infrastructure to train and skill the noncollege bound. In addition, we suggest that small towns should embrace immigration where possible, and to do so in such a way as to integrate non-natives and harness the undoubted economic power that such change can bring. Small towns should also look to invest and develop renewable energy and sustainable agriculture, both of which are natural candidates for being locally crafted and labor-intensive.

States can and should be at the forefront of renewal efforts, and in particular they can stimulate the planning and coordination of regional development, and ease the transition to enhanced cooperation between different bureaucracies. States can also develop best practices in terms of "brain gain" policies, learning from what has succeeded or failed elsewhere. Finally, states can do their part to enforce health and safety laws in agribusiness and industry, and can allocate government stimulus funds to areas most in need.

At the national level, the government should develop a set of policies that can revitalize small towns. This would include targeting stimulus funds directly at stemming hollowing-out by helping towns develop opportunities in renewable energy, sustainable agriculture, and regional tourism. We also recommend that immigration reform should allow longtime undocumented workers a pathway to citizenship, and labor reforms should allow workers in agribusiness to organize to agitate for better wages and safer working conditions. As Sociologists in Action, we have taken our research and recommendations and presented them to the public through various avenues—for example, the Center for Human Potential and Public Policy, and at colleges and universities across the nation. Our book's findings and policy suggestions have also been disseminated widely through reviews in such popular venues as the *Wall Street Journal, Chronicle of Higher Education, Wired.Com, Midwest Book Review,* and *Publisher's Weekly.*

Young adults in Kansas, Minnesota, Upstate New York, Maine, West Virginia, and Wisconsin, to name a few, have written to us to thank us for writing their story. Those who have e-mailed self-identify as achievers, seekers, stayers, and returners, and perhaps one of the best compliments paid to us was when the *Des Moines Register* ran a front-page story about "stayers" in a small Iowa town (Kilen, 2009). At this stage, it is unclear where our journey regarding the hollowing-out problem will end. For now, it feels good to have successfully used our sociological skills to describe what seems to be a universal process, and that people are talking about a problem that has gone unnoticed for so long.

References

Carr, P. J., & Kefalas, M. J. (2009). *Hollowing out the middle: The rural brain drain and what it means for America.* Boston, MA: Beacon Press.

Kilen, M. (2009, December 6). Are overlooked "stayers" keeping rural Iowa alive? *Des Moines Register,* A1.

"WHY DON'T WE DO SOMETHING ABOUT IT?" RESPONSE TO JOB LOSS IN RURAL COMMUNITIES

Leslie Hossfeld

University of North Carolina, Wilmington

Leslie Hossfeld is director of the Department of Public Sociology at the University of North Carolina, Wilmington. She received the 2005 Faculty Fellow Award in Public Policy and Public Engagement at the Institute for Emerging Issues at North Carolina State University. She studies rural poverty and economic restructuring and has made presentations to the U.S. Congress and North Carolina Legislature on job loss and rural economic decline. Dr. Hossfeld is co-chair of the American Sociological Association Task Force on Public Sociology. She works on economic recovery projects for rural North Carolina counties and is cofounder and executive director of the Southeastern North Carolina Food Systems Program.

I first met Mac Legerton, executive director of the Center for Community Action, two months into my new job. At that time, I had just joined the faculty at the University of North Carolina, Pembroke, a small state university in rural Robeson County. It was an uncanny confluence of events—meeting Mac, teaching Social Problems in Robeson County, and what was to follow. I should back up a bit and tell you why this is so unusual: How many of us have the opportunity to teach a section of Social Problems on economic restructuring with over half of the class consisting of the very people we are talking about? Well, this is just what happened to me. I was discussing NAFTA (the North American Free Trade Agreement) and its aftermath, giving a lecture on job loss and what happens when work disappears—when I noticed a real shift in the energy in the classroom. One hand went up, then another, and I began to realize that a good portion of the students in the class were displaced manufacturing textile workers! It seemed an appropriate time for me to keep quiet and learn from the folks who were experiencing this firsthand. It was an eye-opener.

Robeson County is an unusual part of the country: It is the most ethnically diverse rural county in the United States. It is home of the Lumbee Native American tribe (about 38% of the population) and has a large African American population (about 24%) and a rural white population (about 32%), as well as a growing Latino population (U.S. Census Bureau, 2009). It is also a county of persistent poverty (defined by the U.S. Department of Agriculture [USDA] as counties with over 20% of the population in poverty since 1970; USDA—Economic Research Services, 2004). It is also one of the poorest counties in the nation of its size (midsized counties with populations of 65,000 and over), ranking in the top five consistently for many years (DeNaves-Walt, Proctor, & Lee, 2006). The county was heavily hit by the negative repercussions of NAFTA, through which many textile manufacturing firms, in search of lower-cost labor and higher profits, found it more profitable to set up operations outside of the United States, moving first to Mexico and then further overseas. Displaced workers in my class told stories of how their textile manufacturing employers closed up shop and moved away, leaving them without work. Many found themselves jobless after over 30 years with the same company. Some workers described the insult of being sent to Mexico by their employers with the ruse of a "vacation for excellent service," only to learn that they were sent to train future Mexican workers for the very jobs they would lose once they returned from their "vacation"! The stories were infinite, depressing, and heartbreaking. Generations of family members had worked at these firms—the last of the "cradle-to-grave" jobs in the United States. They had mortgages, kids, and parents to care for. And then came NAFTA, the plant closings, American manufacturers racing to Mexico to find the cheapest labor sources they could, and the loss of tens of thousands of jobs in the United States.

Many workers returned to high school to finish their GED. Those who already held high-school diplomas went to college to try to upgrade their skills. However, Trade Adjustment Agreement (TAA) benefits (which provided limited financial assistance for workers to go back to school) quickly ran out, leaving many displaced workers unable to complete degrees, or even to finish their GEDs. The stories of running out of TAA benefits were endless.

Enter Mac Legerton and the Center for Community Action (CCA). Mac and his wife, Donna Chavis, had cofounded CCA over 30 years ago. This remarkable social justice organization had worked tirelessly to tackle the long-standing inequalities related to the critical issues of race, class, and gender that stifled Robeson County. Just as CCA was poised to move ahead with all the key players at the table (Native Americans, African Americans, and rural whites), the bottom dropped out in terms of economic turmoil and job loss. When I met Mac, we clicked immediately—two kindred souls, two like-minded people! Our conversation turned to the high poverty in the area and my students, his community, who were describing firsthand their job loss and difficulties in finding work. Mac turned to me and said, "Why don't we do something about it?" And there and then, a movement was born. "Why don't we do something about it?"—simple words that struck a chord.

Within weeks, we were planning a strategy. First of all, we wanted to document the massive job loss that was unfolding. We knew intuitively what was happening, but we didn't know *empirically* what was going on. We began organizing community members and formed a group of over 70 people from public and private partnerships that became the Jobs for the Future Project. Our goal was to collect data on what was happening around us and to tell our story and get the message out. Mac had worked closely over the years with our congressperson from the region, Mike McIntyre, who was the co-chair of the Congressional Rural Caucus. We decided, fairly quickly, that telling Congress about what was going on due to NAFTA was pretty important. That was our plan: Collect the data and tell the story to Congress.

I began integrating the data collection into two of my classes. Students were excited about the project, all having known someone in the area who had lost a job—and many students were themselves displaced workers. We designed a research project, collecting both qualitative and quantitative data. Students formed committees that focused on organizing the trip to Washington, DC, finding media outlets for the data, and organizing at the grassroots level. CCA worked nonstop to arrange the congressional briefing; conduct outreach throughout the state, nation, and world; find funding for the trip; and pull together displaced workers to share their stories. We partnered with colleagues at neighboring universities who worked with us on data collection and analysis. I received an American Sociological Association CARI grant (Community Action Research Initiative) to support the trip to DC. The University of North Carolina, Pembroke, donated three buses for the trip, and the buses were filled immediately! On March 30, 2004, displaced workers, university administrators, faculty and students, community organizers, researchers, elected officials, journalists, and service providers filled three buses for the overnight trip to DC!

The congressional briefing room was full to the brim! Journalists, congressional staffers, and our three busloads of citizens occupied every inch of the room. I began the briefing, presenting the qualitative and quantitative findings from our research. Our findings reported that Robeson County alone had lost 10,000 manufacturing jobs since the implementation of NAFTA, an estimated ripple effect of $4.5 billion lost in income and tax revenue in a five-county region based on the jobs lost in *one* county. Displaced workers then shared their stories and brought to life the data, giving *testimony* to what happens when work disappears. Elected officials described diminished services, high unemployment, and high crime. One person, who had been selected from our research team to represent the students, provided testimony from the perspective of college students looking for work in an economically devastated community. We also shared in our testimony policy recommendations based on our research findings.

Riding home on the buses, we felt energized and proud. The community had told their story to Congress, and Congress had listened. We wanted to continue the work and continue telling the story so that meaningful social change would take place.

Shortly after the trip to Washington, DC, we organized a national conference on job loss, bringing over 400 people from across the nation to Robeson County to examine the work we had done. Much of our work has been documented in news stories (see http://www2.asanet.org/public/jobless2.html) and on our Web site, www.povertyeast.org/jobs. The Jobs for the Future project received local, national, and international press coverage and, most importantly, our partnerships and networks grew from this project. The policy recommendations we presented to Congress focused on TAA benefit expansion, grant programs for small businesses, and workforce development initiatives.

Though it took several years, components of these policy recommendations have been implemented at various levels. Some of the projects our policy proposals stimulated included North Carolina House Bill 2687 to implement a Poverty Reduction and Economic Recovery Legislative Study Commission, which was implemented in 2008; Local Food initiatives through the Congressional Rural Caucus that were reflected in the 2008 Farm Bill; and Medicaid relief legislative reform for high-poverty North Carolina counties.

While we were proud of the organizing successes, the trip to DC, and the national conference, we knew the work ahead was significant and that much more needed to be done. We used our data to determine our plan of action. For example, our findings indicated that women were the hardest hit by the layoffs because of the gendered nature of textile manufacturing. So CCA created the Women's Economic Equity Project (WEE), a Career Pathways project targeting skills development for rural women workers. Working closely with the growing sectors of health and education, we partnered women with employers in a skills-training and coaching model to help former textile workers translate their manufacturing skills into living-wage jobs in the region. The WEE project includes coaching, peer support, asset development, referral, advocacy, education and training, and need-based scholarship assistance for unemployed and underemployed women in Robeson County. At full capacity, the WEE project will serve 600 women and assist 100 women in moving out of poverty each year.

Our overall focus has been on economic recovery for the region and for rural America in general. We also wanted to approach recovery in as comprehensive a way as possible, and so we began to examine our region sector by sector. Our research in southeastern North Carolina identified the growth sectors of health and education, the challenged sectors of agriculture/forestry and manufacturing, and the emerging sectors of entrepreneurship/small business, and recreation/tourism. Based on these findings, our work has segued to focus on the challenged agriculture sector. We have put forward an economic development initiative around a regional food system that focuses on limited-resource farmers finding market opportunities through institutional buying relationships with our schools, hospitals, colleges, and universities. It is a local food movement that emphasizes buying local, keeping a greater percentage of food dollars within the local economy, and reducing our carbon footprint. While this is but one of many economic development projects that are needed, it provides an avenue toward economic recovery in our region (see www .feastsoutheastnc.org).

The Jobs for the Future project lives on and is embodied in the many community-based projects under the auspices of the Center for Community Action. Sociology is truly the cornerstone for all of our work, providing the tool kit to examine the critical issues affecting our community. We know we have a long road ahead, but with many invested partners and citizens, we know we can bring about meaningful social change.

References

DeNaves-Walt, C., Proctor, B. D., & Lee, C. H. (2006). *Income, poverty, and health insurance coverage in the United States: 2005.* P60-231. Washington, DC: U.S. Department of Commerce, U.S. Census Breau. Available at http://www.census.gov/prod/2006pubs/p60-231.pdf

U.S. Census Bureau. (2009). *State and county QuickFacts: Robeson County, North Carolina.* Washington, DC: U.S. Department of Commerce. Available at http://quickfacts.census.gov/qfd/states/37/37155.html

USDA—Economic Research Services. (2004). *Rural income, poverty, and welfare.* Retrieved from http://www.ers.usda.gov/Briefing/IncomePovertyWelfare/

DISCUSSION QUESTIONS

1. Yvonne Vissing states that, "while having a good heart is important . . . one also needs a good head in order to create positive social change." How does her work exemplify this fact?

2. As Vissing points out, sociologists understand that social institutions influence one another. Describe some ways the interactions among (a) the family, (b) the economy, and (c) the government impact children.

3. Can you relate to any of the three groups of young people that Carr and Kefalas describe in their piece (the achievers, seekers, and stayers). If yes, how so? If not, why not? Do you think that the town where your college resides (or your hometown) has similar categories of young people? Why or why not?

4. If a mayor of a rural middle-American town asked you to use your sociological imagination to help her come up with means to address the "hollowing-out" of her town, how would you advise that mayor?

5. Hossfeld turned to the government (Congress) for help in addressing the rampant unemployment in her area. Do you think the federal, state, or local government is capable of impacting our economic institution? If no, why not? If yes, do you think each respective level of government (federal, state, local) is doing enough to combat unemployment in your area? What makes you think so?

6. Hossfeld used her sociological imagination to relate her students' personal troubles (being out of work) to a public issue (the impact of NAFTA). Think of a personal trouble that is affecting a group of people on your campus or in the local community that you can relate to a larger public issue. How does your means of addressing the problem change when you think of it as a public issue rather than a personal trouble?

7. Think about your own family. How does the economic situation of your family impact how you relate to one another? Do you think you and your other family members would interact differently if your family was in a (a) higher or (b) lower social class? How might your relationships be impacted if the primary breadwinner(s) no longer provided the most money for the family? Now think about the family and the economy at the structural level of social institutions. In what ways do you think the social institutions of the family and economy impact one another?

8. Imagine you are a Sociologist in Action studying work–family balance in a corporation. What social patterns might you expect to see? How would you set up a study that would reveal if those patterns do exist and their impact on (a) the economic output of the individuals in your study and (b) the families of the individuals in your study?

RESOURCES

The following Web sites will help you to further explore the topics discussed in this chapter:

ASA Section on Economic Sociology	http://www2.asanet.org/sectionecon/
ASA Family Section	http://www2.asanet.org/sectionfamily/
Free the Children	http://www.freethechildren.com/what wedo/international/
Global Issues: World Hunger and Poverty	http://www.globalissues.org/issue/6/ world-hunger-and-poverty
Marriage and Family Processes	http://www.trinity.edu/~mkearl/family.html
Sociosite Economic Sociology	http://www.sociosite.net/topics/econsoc.php
Sociosite Family and Children	http://sociosite.net/topics/familychild.php
United for a Fair Economy	http://www.faireconomy.org

To find more resources on the topics covered in this chapter, please go to the Sociologists in Action Web site at **www.sagepub.com/korgensia2e.**

Chapter 14

Social Institutions, Continued (Education, Government, Religion)

In this final chapter, we look at how some Sociologists in Action are examining the other three major social institutions not covered in Chapter 13: education, government, and religion. There are myriad ways that sociologists can study these institutions and use their findings to promote a more just society. The sociologists in this chapter share how they have been able to recognize social patterns within these respective institutions and suggest ways to improve them.

Sigal Alon opens the chapter with "A Sociologist as a Social Seismographer: Understanding the Earthquake in Class Inequality in U.S. Higher Education." In this piece, she explains how she felt compelled to make sense of patterns indicating a divergence in standardized test scores over the years between lower- and upper-class students applying to colleges. Using her sociological eye, she sifted through the data and noticed that the test scores of the wealthier students rose in tandem with increased college admission standards. Alon argues that, as entry into college became more competitive, class advantages became more important as wealthier students gained more of those coveted admission slots. These findings have led her to advocate for a move to class-based affirmative action in college admissions in order to mitigate the disadvantages facing students who have fewer economic resources.

Dadit Hidayat, Randy Stoeker, and Heather Gates reveal how institutions of higher education "can use their knowledge and resources to support community action." Through "Promoting Community Environmental Sustainability: Using a Project-Based Approach," they show how a university–community partnership using a project-based approach "addresses environmental sustainability issues *and* creates social change for all involved." The university courses are designed around the project work to "prepare future leaders in the field of environmental sustainability by providing professional development experience." They also help environmental activists use sociological tools to "better target and focus their work."

In "Using Sociology to Counter Stereotypes: The Case of American Muslims," John O'Brien and Besheer Mohamed describe how they reacted to the prevalence of negative stereotypes about Muslim Americans after September 11, 2001. As Muslims, they were frustrated. As sociologists, it made them "wonder what [they] could do about it." Both conducted carefully designed research projects on Muslim Americans. "What emerged was a picture of Muslim Americans as a diverse group, concerned with the same issues as most ordinary people, rather than extremist ideology or religious fundamentalism." They are now using various strategies to disseminate their findings to the public in order to "counter inaccurate stereotypes" of Muslims. They both "believe that rigorous research has the potential to illuminate hidden social truths and, in so doing, spur positive social change."

In "Democracy Matters: Giving Students a Political Voice," Joan Mandle describes how she and her adopted son, Adonal Foyle, joined forces to make a positive impact on society. When Foyle was in college, he was concerned that his friends' potential power to promote change was negated because students on his campus were either divided among various causes or cynical and apathetic about the possibilities of political action. When Foyle became the eighth pick in the 1997 NBA draft, mother and son suddenly had the financial means to take a giant step to address this problem and used their resources to develop Democracy Matters (DM). Mandle describes the work of DM, a nonpartisan student organization they created "to harness the collective power of young people, and make their voices heard politically so that they could become effective social change advocates for themselves and for others." Through DM, thousands of college students on over 500 campuses have taken part in a social movement to make our government more democratic.

In the final piece, "Out of the Tower and Into the Capitol: How Sociology Students Helped Spark the Wisconsin Uprising," Charity Schmidt explains how she and her fellow sociology teaching assistants at the University of Wisconsin used their sociological training, as they and hundreds of thousands of other Wisconsin residents attempted to influence their state government during the 2011 Wisconsin Uprising. She vividly illustrates how "the experience of thousands of Americans together, demanding and exercising their democratic rights, forever transformed all those involved" in the occupation of the Wisconsin State Capitol.

A SOCIOLOGIST AS A SOCIAL SEISMOGRAPHER: UNDERSTANDING THE EARTHQUAKE IN CLASS INEQUALITY IN U.S. HIGHER EDUCATION

Sigal Alon

Tel Aviv University, Tel Aviv, Israel

Sigal Alon is a senior lecturer in the Department of Sociology, Tel Aviv University. She studies the U.S. postsecondary education system to comprehend the mechanisms underlying race and class disparities in access, experiences, and performance of students at selective and nonselective institutions. She also examines the social implications of affirmative action and financial aid policies in postsecondary education. She is now studying questions related to equality of educational opportunity at Israeli flagship universities. Dr. Alon has published articles in several leading journals, including *American Sociological Review, Social Forces, Work and Occupations, Sociology of Education, Economics of Education Review,* and *Social Science Research.*

Several years ago, while I was running statistical analyses on three cohorts of high-school graduates (1972, 1982, and 1992) for a new study that I planned to pursue, I stumbled across a finding that captured my attention. I found that, while during the 1970s the gap in the standardized test scores of high-school graduates between the top and bottom socioeconomic quartiles narrowed, it widened during the 1980s. Basically, since the mid-1980s, students from the most privileged class, those with the most economic and other resources, improved their test scores, while the underprivileged, those with the least economic and other resources, were unable to follow suit. This struck me: How and why could such a dramatic change have occurred over the course of just one decade? I decided to abandon my original research question and headed directly into trying to unravel this puzzle.[1]

---

[1]The research findings were published as the following: Sigal Alon, "The Evolution of Class Inequality in Higher Education: Competition, Exclusion, and Adaptation" (*American Sociological Review,* 74, 2009), 731–755.

My desire to understand the underlying causes of the widening gap went beyond my interest in the changing test score distributions themselves. Rather, it was rooted in my belief that this increasing polarization (the concentration of the privileged class at the top of the test score distribution and the underprivileged at the bottom) had serious implications for socioeconomic class–based inequality in higher education (hereafter in this article, I will refer to these as *class inequality* or the *class divide*). Furthermore, since test scores are important determinants of access to higher education, which in turn is one of the most significant factors that affects future socioeconomic status, a widening test score gap between the most and least privileged may be an ominous indicator of widening social inequality in general.

Sociologist Charles Tilly, in his book *Durable Inequality* (1998), describes the work of analysts of social inequality as similar to that of seismographers (earthquake analysts), an analogy that I find fitting. In the context of this project, I envisioned that the shifting of plates beneath the earth's surface (i.e., a widening test score gap) may have created an earthquake on the surface (i.e., increasing class inequality in higher education). The first step, therefore, involved the determination of whether a surface earthquake in higher education had actually occurred during the period in question in accordance with the changes in the test score gap. If not, it would imply that the widening gap was of little consequence. If so, however, I would be left to decipher the causes of this sudden shifting of plates.

Phase I: Detecting the Earthquakes in Class Inequality in Higher Education

Given that class inequality exists from kindergarten through high school, inequality between the haves and have-nots in access to postsecondary institutions, especially to the most selective schools, would seem obvious and ubiquitous. Indeed, numerous studies have concluded that inequality in higher education is persistent over time. However, what does "persistent" really mean in this context? Does it mean that the level of inequality is stable over time? The changes in the test score gap in recent decades seemed to indicate otherwise.

My findings confirmed this suspicion. I discovered that the magnitude of the class divide in higher education is tightly bound to the magnitude of the gap in test scores. When the gap in test scores was small, as was the case from the mid-1970s to the mid-1980s, inequality in all postsecondary institutions declined. However, statistical analysis of the postsecondary institutions of 1982 and 1992 high-school graduates showed a widening of the class divide. This means that the class divide during the late 1980s was larger a decade later. That is, underprivileged high-school graduates in the late 1970s and early 1980s had better chances of getting into a four-year college and of being admitted to more selective schools than their peers a decade later! In sum, my findings demonstrated that the polarization in standardized test scores had facilitated an ensuing "earthquake" in the mid-1980s: Class inequality in higher education, which until that point had been in decline, suddenly started to rise. But why? What led to the widening gap in test scores?

Phase II: The College Admissions Market—Competition and Selection

To understand what stood behind the changes in test score distributions, I embarked on an investigation of the level of competition in college admissions. My research looked into both the supply of slots in and the demand for postsecondary education. My hypothesis was that there are periods when the level of competition in admissions is low (supply of seats surpasses demand); admission barriers are relaxed; and the postsecondary education system becomes more inclusive, especially at the lower rungs of the ladder. Other times, when competition is high, the increasing demand for a college diploma intensifies the college squeeze, and admissions officers are able to select students from a surplus of high-quality applicants.

In my quest to understand the trends in supply and demand in higher education, I conducted extensive desk research. In order to obtain a broad historical perspective, I compiled data beginning from the 1950s. I gathered information from various online sources (such as the Census Bureau, the U.S. Department of Education, and the National Center for Education Statistics) and found supplementary information in the archives of several selective institutions, including data on the applicant and admit pools and the annual admission rates. Some of the information was available online, while the rest was provided by the schools' institutional research offices upon my request.

As I sifted through the data I had gathered, several phases in college admissions were revealed before and during the period of my review of test scores. The first phase started in the mid-1950s with the massive expansion of the postsecondary system. Starting in 1955, applications and enrollment swelled at colleges and universities across the United States. To accommodate the steep rise in demand, the postsecondary system expanded dramatically by increasing the capacity of existing institutions and adding new institutions, such as community colleges, regional universities, and additional branches of existing universities. The second phase started in the late 1960s and early 1970s. Several social and economic factors lowered the demand for higher education in the midst of an ongoing expansion that continued to add new seats every year. As a result, the supply of seats in most institutions surpassed demand, which led to lower levels of competition for the available seats.

Finally, my historical review indicated that yet another phase began in the mid-1980s. As the rate of college attendance increased, the competition for slots in higher education increased dramatically, especially at selective institutions. While this process was gradual and had started before the 1980s, it took off as the competition increased.

Taken together, the findings suggest that when the competition is low, institutions lower their test score admission thresholds. Conversely, when the competition in admissions is fierce, the selection process is tight. How does the population of high-school graduates respond to these changes in the level of competition and admission practices? Does class-based convergence (the process by which the gap in test scores between the bottom and top section is narrowing) or polarization in the distribution of test scores reflect their behavior? How does their response shape the magnitude of the class divide in access to higher education?

Phase III: Developing a Comprehensive Theory Regarding the Evolution of Class Inequality in Higher Education

My quest had been ignited by a graph that pointed to a widening test score gap between the most and least socioeconomically privileged groups since the mid-1980s. I argue that the improved position of the privileged in the distribution of standardized test scores reflects their *adaptation* to the rising value of test scores in ensuring admissibility. The privileged, aware of the high level of competition in admission and the rising emphasis on test scores in admission decisions, undertook the required behavioral adjustments, while the less privileged were unable to follow suit. The privileged classes have a better understanding of the postsecondary educational landscape and competitive admissions process, and thus, they invest more heavily in resources that promote college attendance. For instance, vigorous use of expensive test-preparation tools, such as private classes and tutors, is a key element in their adaptation strategy. This preparation was likely a significant factor in boosting test scores, while less privileged youth, on the other hand, did not have the knowledgeable and proactive parents, the experienced counselors, or the resources to finance private tutors. In turn, this led to a class-based polarization of resources, placing privileged students in a better starting position for admissions and widening the class divide.

In contrast, when competition in admissions is low, as was the case during the 1970s, the importance of test scores for college admission diminishes and a narrowing of the class divide ensues.

Phase IV: Implications for Public Policy

Building on these findings, I argue that if we wish to broaden disadvantaged students' access to four-year and selective institutions, we must consider class-based affirmative action policies. Giving an edge in admissions to low–socioeconomic status (SES) seniors will merely compensate for the competitive advantages that accrue to the privileged through adaptation. Those most damaged by adaptation—talented underprivileged seniors— would benefit the most from a policy that cultivates dreams, aspirations, and ambitions for a type of education that, without preferential treatment, is beyond their reach.

Through publishing my work in a prominent American Sociological Association (ASA) journal that is reviewed by leading commentators, journalists, and other newspeople, I was able to put out my policy recommendations to the public sphere. Although my findings were published just a few months ago (as I write these words), this recommendation has already stirred a public debate regarding the need, the feasibility, and the design of class-based affirmative action policy in the United States as well as in other countries (see links to media coverage of the study on the *Sociologists in Action* Web site). This public conversation about higher education and social change may eventually lead to the implementation of these policy recommendations by policy makers in higher education and the government.

It is thrilling for me that my academic work can have a tangible influence on people's lives. It is also fascinating to closely monitor the development of what is usually an elusive concept: a "social change."

References

Tilly, C. (1998). *Durable inequality.* Berkeley: University of California Press.

PROMOTING COMMUNITY ENVIRONMENTAL SUSTAINABILITY USING A PROJECT-BASED APPROACH

Dadit Hidayat

University of Wisconsin

Randy Stoecker

University of Wisconsin

Heather Gates

The Natural Step Monona

Dadit Hidayat is a PhD student in the Nelson Institute for Environmental Studies at the University of Wisconsin. Dadit has volunteered with The Natural Step Monona (TNS Monona) since fall 2008, which is largely driven by his interest in addressing a serious gap between the academic and local community in how to address environmental problems.

Randy Stoecker is professor in the Department of Community and Environmental Sociology at the University of Wisconsin, with a joint appointment in the Center for Community and Economic Development. Randy is continuing to work on amplifying the community voice in service learning and providing research support for nonprofit organizations.

Heather Gates is director and cofounder of The Natural Step Monona, a grassroots organization educating about, advocating for, and promoting sustainability. A catalyst for change, the group helps move the city, its residents, and people throughout the region toward a more sustainable future. Gates is a Nelson Institute Community Fellow.

Environmental sustainability is a topic of discussion across the globe. But getting people to act collectively on environmental sustainability, especially in local communities, is fraught with challenges. Sociologically, the challenge is one of understanding how collective action works in order to mobilize community members around environmental issues. An important sociological aspect of mobilizing people for collective action is *framing* those issues effectively (see Benford and Snow, 2000). Frames are toolboxes of interpretation that help us make sense of the world. We have frames of not just what is right and wrong but even about what does or does not exist. For example, people have various frames of what *sustainable* is. One person might interpret ethanol as sustainable, comparing it to fossil fuels, while another might interpret it as unsustainable based on an analysis of the energy required to grow and transform the corn into ethanol. Activists weigh into these controversies, attempting to bridge their frames with community members' frames or transform community members' frames to fit with the activists' frames.

Some people may not be aware of important environmental problems, such as water pollution. Knowing that community members likely care about their children's health, an environmental activist can offer that water pollution is endangering their children's health, reframing drinking water pollution as a children's health issue. A related challenge, then, is finding ways to connect highly technical scientific knowledge with community members' experiences of environmental issues in ways that could inform practical action.

Community-based organizations working on environmental issues need to organize people to act collectively, and that often requires changing people's frames to convince them to get involved. However, environmental researchers generally have neither adequate training in science communication (Meredith 2010, Schmidt 2009) nor the necessary skills to organize civic engagement based on good research. Universities and colleges can use their knowledge and resources to support community action. Higher education institutions typically use one of three strategies, but each has weaknesses.

Translational research (Mercher et al., 2007) tries to "translate" scientific research to lay audiences but doesn't ask the community what research they want. *Participatory action research* usually involves researchers and community members working together (Ballard & Belsky, 2010) and does ask community members what research they want, but it too often assumes that doing the research is more important than acting on the findings. *Service-learning*—students working directly with communities—focuses much more on student learning than community change, and thus also falls short of being an effective strategy for achieving a sustainable planet (Stoecker & Tryon, 2009).

We propose a fourth strategy—*project-based research*—for a university–community partnership that addresses environmental sustainability issues *and* creates social change for all involved. In project-based research, researchers and service-learners collaborate with community members to *diagnose* community issues, *prescribe* solutions for those issues, *implement* the prescriptions, and *evaluate* the outcomes (Stoecker, 2005). The collaboration is based on the three Cs—*commitment, communication, and compatibility* (Hidayat et al., 2009). This new strategy combines all of the strengths of the other three strategies, while ultimately moving toward realizing actual and viable change.

After participating in a community-based research project as a graduate student with Randy Stoecker, Dadit Hidayat became fascinated with conducting community-based research. Interested in building mutual trust and partnership with a community, he sought out Heather Gates in 2008. Heather leads The Natural Step Monona (TNS Monona), a grassroots community organization that passionately works to promote sustainability in Monona, a small city across the lake from Madison, Wisconsin.

Dadit volunteered with TNS Monona and saw how a university–community partnership could support the organization's work. Learning of a funding opportunity from the Community and Environmental Scholars Program (CESP)[2], the three of us collaborated to secure the funding and develop two sequential capstone courses for undergraduate students in the environmental studies certificate program. Organized around the project-based model, we designed the courses to meet TNS Monona's need to gauge their effectiveness and better engage the community in sustainability.

Course 1—Spring 2011: Community-based Research with The Natural Step Monona (TNS Monona)

Sociological research tools formed the centerpiece for our first course. In planning this course, Heather hoped that students could discover how TNS Monona and the City of Monona's sustainability efforts have affected community members' sustainability practices and belief systems—their environmental frames—while spreading the word about TNS Monona. The research combined community-based evaluation with a diagnosis of what environmental sustainability issues Mononans cared about. The City of Monona Sustainability Committee and the City Administrator provided community meeting space and publicly endorsed our project, adding credibility to the effort.

Twelve students (the maximum allowed) signed up for the first capstone class. After a few weeks of in-class planning, we decided that the best method for the evaluation and diagnosis would be a community-wide survey. The students spent four weeks collaboratively developing the survey with TNS Monona. Then, with the capstone students, 14 TNS Monona

---

[2]CESP is administered by the Nelson Institute for Environmental Studies at the University of Wisconsin-Madison. The program trains undergraduates to work with community-based environmental organizations and gives undergraduates a place to discuss the links between Environmental Studies and community service.

volunteers, and a civic engagement campus expert, we canvassed for four afternoons during late March and early April 2011. We knocked on nearly 3,100 doors in Monona, distributing paper surveys. We received just over 600 responses, which was quite successful considering the absence of any follow-ups.

Partnering with the city, we organized a public meeting in May to present the survey results and engage Mononans in a discussion about what the results meant for their community. Forty-four residents, including the newly elected mayor and two local media representatives, attended. As residents discussed the survey results, they considered how TNS Monona and the city should act on the findings. Residents learned about the City of Monona's sustainability efforts, and the frames we used helped residents see TNS Monona and the City of Monona as collaborators in local sustainability efforts. The local newspaper reported on the event, framing the survey results as emphasizing how strongly Mononans felt about environmental issues, such as clean drinking water and clean lakes.

By the time of this meeting, we already knew that we would have a second capstone course in the fall. So the survey results and the community discussion produced by this first capstone course provided the foundation for TNS Monona to create a grand plan that our second capstone course would support.

Course 2—Fall 2011: Community Organizing with TNS Monona

The students in the first course had discovered what Mononans thought about sustainability, what, if anything, they were doing about it, how they perceived TNS Monona's efforts so far, and what their greatest community concerns were. Next, we wanted to engage the residents in collective action that would address these concerns. An important finding from the survey was that clean lakes and clean drinking water were the top two community issues identified by Monona residents. TNS Monona realized the opportunity to use the concerns over water as a leverage point to build long-term engagement and action for sustainability. Focusing on water for the near term, TNS Monona successfully influenced the mayor to proclaim 2012 as the "Year of Water," and organized a year-long Water Conservation Challenge in 2012 with households competing to reduce water usage (incentivized not only with prizes, but by water rates going up 32% in November 2011).

Meeting several times in summer 2011, we agreed on two goals for the fall capstone course: to make Monona's use and handling of water more sustainable, and to make TNS Monona's voice louder. We then designed the course to engage local community-based organizations (CBOs) in generating ideas for their own Year of Water projects.

After eight weeks of education on water and related sustainability issues, students in the second course prepared a PowerPoint presentation on water conservation, developed scripts for contacting CBOs, and received training on how to recruit groups to take on community projects supporting the Year of Water. Once again employing collective action framing techniques, we encouraged community groups to think about easy and manageable water-related projects. In November and December, students contacted 54 school and community groups, and gave presentations on water sustainability to 12 of them. The students also created a brochure and a web page for communicating information about the Year of Water.

The students obtained pledges from a handful of groups and wrote reports on the other groups they contacted. TNS Monona then followed up with those groups, and contacted others, to gain public commitments from a total of 11 groups. The pledges are diverse. On Earth Day, Saint Stephen's Lutheran Church hosted a workshop on how to construct and install rain barrels and displayed easy ways to conserve water in the home. The City of Monona Sustainability Committee is already making good on a pledge to establish an online system where residents can compare their water efficiency to that of others (accomplished with another service-learning class from Marquette University). 4Pillars4Health EcoSpace is organizing a Water Walk around Lake Monona to be led by First Nations Ojibwe Grandmother Josephine Mandamin and William Waterway Marks. Friends of the Monona Senior Center organized two screenings of the film *Waterlife*. Monona Grove Liberal Arts Charter School for the 21st Century (MG21) High School has conducted tests of area water bodies. And the Monona Garden Club will assist the Winnequah Children's Garden in developing a rainwater harvesting system.

What will come of all this? We are not yet completely certain, as most of these projects are just getting underway. But we now have a third capstone course planned for the spring of 2013, and those students will work with TNS Monona to do research so that we can find out how much has changed, and what is left to change. TNS Monona will then have knowledge to help it plan its next steps to building a more sustainable community in Monona.

Lessons Learned

One challenge of much university–community engagement is that it is sporadic and short-term, making it very difficult to show significant impact.

These back-to-back capstone courses (spring 2011 and fall 2011) offered the opportunity for a longer-term commitment that went from the initial evaluation and needs assessment through the application of a strategy. The project-based model gave us time for logical reasoning based on our survey results, leading us to develop sustainability plans for the community in a strategic way. We also crafted messages that resonated with the community's concerns, using water to loosen the doors of communication and engage others in action.

These courses helped CESP prepare future leaders in the field of environmental sustainability by providing professional development experience. They also helped students and TNS Monona think more sociologically about environmental sustainability in two ways. First, the survey helped show the value of sociological research for illuminating what opportunities might exist for *collective action,* rather than focusing just on an individual behavior change approach. And, second, the collective action framing strategies helped engage groups and networks of people in implementing the Year of Water in Monona to benefit the entire community. For the three of us, we have realized the importance of our own collaboration. Sociology, to be useful, requires activists to put sociological knowledge into action in the real world. And activists, when they have access to sociological knowledge, can better target and focus their work.

References

Ballard, H. L., & Belsky, J. M. (2010, October-December). Participatory action research and environmental learning: Implications for resilient forests and communities. *Environmental Education Research, 16*(5–6), 611–627.

Benford, R. D., & Snow, D. N. (2000). Framing processes and social movements: An overview and assessment. *Annual Review of Sociology 26,* 611-639.

Mercer, S.L., DeVinney, B. J., Fine, L. J., Green, L. W., & Dougherty, D. (2007, August). Study designs for effectiveness and translation research: Identifying trade-offs. *American Journal of Preventive Medicine, 33*(2), 139–154.

Meredith, D. (2010, May 16). Please explain: Training scientists to be better communicators. *The Chronicle of Higher Education.* Retrieved from http://chronicle.com/article/Please-Explain-Training/65560/

Schmidt, C. W. (2009). Communication gap: The disconnect between what scientists say and what the public hears. *Environmental Health Perspectives, 117,* A548–A551.

Stoecker, R. (2005). *Research methods for community change.* Thousand Oaks, CA: Sage.

Stoecker, R., & Tryon, E. (Eds.). (2009). The unheard voices: Community organizations and service learning. Philadelphia, PA: Temple University Press.

USING SOCIOLOGY TO COUNTER STEREOTYPES: THE CASE OF AMERICAN MUSLIMS

John O'Brien

University of California, Los Angeles

Besheer Mohamed

Pew Forum on Religion and Public Life

John O'Brien is a PhD candidate in sociology at UCLA. His research uses ethnographic methods to study the workings of culture, religion, race, and identity at the level of everyday social life. His dissertation project explores the daily lives and cultural dilemmas of a group of Muslim American youth growing up in an American city. An article from this project was published in *Social Psychology Quarterly*.

Besheer Mohamed received his PhD from the Department of Sociology at the University of Chicago. His dissertation examines the relationship between religious identity and social attitudes among American Muslims. He also earned a master's degree in Middle East Studies from the University of Chicago.

Think honestly for a moment about the following: What do you think of when you hear the word *Muslim*? How do you expect a Muslim to look, think, or behave? The set of appearances, expectations, and beliefs about Muslims that came to your mind—whether positive or negative—is probably an example of a "stereotype." As sociologists and social psychologists define it, a *stereotype* is a mental representation of a social group. Stereotypes are part of how our mind makes sense of the social world, by categorizing sets of people and associating them with certain traits, appearances, and behaviors (Mackie, Hamilton, Susskind, & Rosselli, 1996). Stereotypes are not inherently harmful, but they have the potential to lead to unfair, discriminatory, and even destructive treatment of members of certain social groups (Major & O'Brien, 2005). Stereotypes are especially dangerous when they include misinformation and inaccurate portrayals of members of a given race, ethnicity, religion, nation, gender, or other social identity. These inaccurate stereotypes can spread and increase the likelihood of personal harassment, discriminatory government policies, or even international conflict.

As American Muslims, we were particularly aware of the significant increase in the negative stereotyping of Muslims by non-Muslims after the attacks of September 11, 2001, both in the media and in everyday life. We heard people say things that we knew to be untrue, such as: "Most Muslims supported the attacks on the World Trade Center," "Muslims are violent people," and "All Muslim women are mistreated and oppressed." Part of what makes inaccurate stereotypes like these so dangerous is that they include the assumption that all or most members of the stereotyped group think or act in the same way. Even if a few Muslims are violent and support terrorist attacks, this certainly does not mean that all or most do. And it does not mean that the people who acted this way did so *because* they were Muslim (Brubaker, Feischmidt, Fox, & Grancea, 2007). As Muslims, the prevalence of such negative stereotyping left us frustrated. As sociologists, it made us wonder what we could do about it.

Because sociology is a science devoted to studying the social world, it provides powerful tools for countering inaccurate stereotypes. Sociological research on the actual attitudes and behaviors of members of social groups can be used to challenge simplified portrayals of these groups. For example, in his book *Sidewalk*, sociologist Mitch Duneier (1999) used ethnographic methods to discover that homeless street vendors, rather than being deviant and disorganized, were working quite hard to make a respectable living in one of the only ways they could. This is a case in which sociology was used to dispel an inaccurate stereotype of a social group by presenting an alternate, more accurate "summary image" of that group (Katz, 1997).

In order to counter inaccurate stereotypes of Muslims—and specifically Muslim Americans—we have used sociological methods to find out what American Muslims are *really* like, and then we worked to share this information as widely as possible. Each of us used his own methodological approach to study the lives, beliefs, attitudes, and behaviors of Muslim Americans, and to investigate the truth behind the stereotypes.

Besheer used both quantitative and qualitative methods for his most recent research project. He used the quantitative method of survey analysis in order to investigate how religious identity is related to social attitudes among American Muslims. A quantitative approach was useful for this part of the project, because it allowed him to look at data from a large number of Muslim Americans across all ages, races, and walks of life. Through this analysis, he found that whether or not a Muslim feels religion is important, it has very little to do with his or her views on most political or social issues. For example, how traditional a Muslim's religious beliefs are cannot be used to predict personal views on al-Qaeda, immigration, or nearly any of the 14 social issues that he tested.

Besheer also used the qualitative method of in-depth, one-on-one interviews. A qualitative approach allows for the collection of detailed data from each person, and thus provides a nuance often missing from solely quantitative studies. These interviews uncovered the fact that, though there is not a strong relationship between religious beliefs and social attitudes, American Muslims do draw on their faith for guidance on issues as diverse and mundane as recycling and healthcare reform. Taken together, the quantitative and qualitative findings show that American Muslims are not striving to impose, or even live by, a dogmatic version of Islamic *shariah* law. Rather, they see their faith as a source of strength and encouragement to go beyond personal desires and look toward the common good, just as many other Americans do.

John used the qualitative method of ethnographic observation to conduct a three-and-a-half year study of a group of Muslim American teenagers growing up in an American city. Each week, he spent between 5 and 10 hours with the same group of young Muslims, going to the mosque, hanging out on the street and at their homes, or driving around in cars. An ethnographic approach was useful for this project because it allowed John to observe Muslim youth in their daily lives and routines. In these settings, the youth could act naturally, and did not think of themselves primarily as subjects of research. Nor did they feel pressure to give the "right" answers or act the "right" way. Of course, John did explain from the beginning that he was a researcher, but over time the teenagers came to trust him and sometimes act as if he was not even there.

As a result of analyzing the over 1500 pages of typed field notes on his observations, John discovered that—far from being violent extremists—these Muslim youth were mainly concerned with the issues and activities that most American teenagers deal with: school, money, music, socializing with friends, and cars. Although being a religious Muslim *was* important to the youth, it did not dramatically conflict with or overwhelm their identity as an American teenager. This discovery countered the mainstream stereotype of Muslims, and religious people more generally, as primarily driven by religious or ideological concerns. Instead, John found that these youth were basically run-of-the-mill teenagers, albeit with some special concerns related to balancing their faith with American youth culture and responding to occasional harassment.

Neither Besheer's nor John's findings—both rooted in social scientific methods of empirical research—provided evidence to support the prevalent stereotypes of Muslims as violent or driven primarily by religious ideology. Instead, what emerged was a picture of Muslim Americans as a diverse group, concerned with the same issues as most ordinary people, rather than extremist ideology or religious fundamentalism. This does not mean that there are no Muslims who do support or participate in these activities and beliefs, but our research suggests that it is not accurate to say that *in general* Muslims think or act this way.

Research does not, however, counter stereotypes by itself. In sociology, as in most sciences, there is a large gap between the state of scholarship and the level of knowledge of the lay public. In order to help the public make decisions based on social science research rather than the inaccurate stereotypes permeating our culture, we have pursued three different strategies, one together and two separately. First, we used our research findings to write a joint article countering the stereotypes of Muslims as supportive of terrorism in the magazine *Contexts* (Mohamed & O'Brien, 2011). Second, John aims to turn his ethnographic study into a published book, which can be read by undergraduates as well as wider audiences seeking to understand the daily lives and concerns of Muslim American youth. Third, Besheer has joined the staff of the Pew Research Center's Forum on Religion and Public Life. He hopes that his research and work there will contribute to the Pew Forum's mission to promote a deeper understanding of issues at the intersection of religion and public affairs for political leaders, journalists, scholars and citizens.

While these efforts at research and dissemination will not singlehandedly end inaccurate stereotyping of Muslims, they may make small changes in the way people think, which can have rippling effects. We have already seen some evidence of positive outcomes from our work. In sharing his research with colleagues, friends, and family, John has repeatedly heard people remark, with some surprise, "Oh, these are just like normal kids." Although it is frustrating to think that Muslim youth would not be seen as normal in the first place, the fact that empirical research can shift peoples' perception of others—even in small ways—gives us hope in the power of sociology to counter inaccurate stereotypes. This is one reason why we continue to work as sociologists—because we believe that rigorous research has the potential to illuminate hidden social truths and, in so doing, spur positive social change.

References

Brubaker, R., Feischmidt, M., Fox, J., & Grancea, L. (2006). *Nationalist politics and everyday ethnicity in a Transylvanian town*. Princeton, NJ: Princeton University Press.

Katz, J. (1997). Ethnography's warrants. *Sociological Methods and Research 25*(4), 391–423.

Mackie, D. M., Hamilton, D., Susskind, J., & Rosselli, F. (1996). Social psychological foundations of stereotype formation. In C. N. Macrae, C. Stangor, & M. Hewstone (Eds.), *Stereotypes and stereotyping*, pp. 41–78. New York, NY: Guilford Press.

Major, B., & O'Brien, L. T. (2005). The social psychology of stigma. *Annual Review of Psychology, 56*, 393–421.

Mohamed, B., & O'Brien, J. (2011). Ground zero of misunderstanding. *Contexts, 10*(1), 62–4.

OUT OF THE TOWER AND INTO THE CAPITOL: HOW SOCIOLOGY STUDENTS HELPED SPARK THE WISCONSIN UPRISING

Charity A. Schmidt

University of Wisconsin, Madison

Charity A. Schmidt is a PhD student in Community & Environmental Sociology at the University of Wisconsin-Madison with a UW masters in Latin American Studies and a bachelors from the University of Michigan. She teaches about race and ethnicity at Madison College and is a project assistant at the Center on Wisconsin Strategy (COWS). Her work focuses on housing and land use in urban spaces. She is active in the UW-Madison Teaching Assistants' Association (TAA), and was elected co-president in 2012. She coordinates with a wide range of community and campus organizations in the movement for economic and social justice.

It's not surprising that students were so involved in the Wisconsin Uprising in 2011. In Madison, the state capitol and the flagship university are joined by a pedestrian mall, providing both a symbolic and physical connection between state and students (not to mention the perfect march route). Moreover, the year marked the 100th anniversary of the Wisconsin Idea, when university professors helped elected officials generate a round of legislation that ushered in an era of progressive politics and a commitment by the university that still lives today—to improve people's lives beyond the classroom and engage with the people of the state to build their capacity for self-governance. This mission was exemplified when we occupied our state capitol for 17 nights in February and March of 2011.

On February 11th, Governor Scott Walker introduced his Budget Repair Bill (later known as Act 10), which put the burden of solving the budget deficit on the backs of public employees and the poor, a deficit exacerbated by generous tax breaks to corporations and the wealthy. The bill stripped unions of their rights, including the right to collectively bargain over benefits and working conditions, a violation of Wisconsin's tradition of progressive labor standards. The Bill also slashed health care for seniors and low-income families, and gutted funding for public education at all levels. The state erupted, and public employees started organizing. K–12 teachers called in sick forcing schools to close, the UW-Madison Teaching Assistants' Association (TAA, the first ever union of graduate student workers) called for campuswide teach-outs, and students across Wisconsin flocked to their Capitol.

My sociology department at the UW had the largest representation of graduate students spending day and night at the Capitol. So, what made so many of us spend every waking moment supporting the occupation? Sociologists are fascinated with processes of social change; we strive to understand how and why people struggle to improve their lives and bring justice to their social world. The sociological perspective accounts for the history of human struggle, allows us to understand the intricate social patterns of the present and compels us to care about the society we are always creating. It reveals that being a member of society is never a passive act.

Foreseeing budget cuts from a new governor elected with Tea Party support, the TAA had serendipitously planned an action for Valentine's Day. Our fears were confirmed by Governor Walker's bill just days before, which kicked off an attack on UW that would result in $300 million in budget cuts, tuition increases, and "flexibility" for administration to act as a private university. Wisconsinites recognized the impending threat to higher education, and on February 14th, 1,000 students and community members dumped thousands of valentine cards on the governor's desk, which read "We ♡ U W. Governor Walker, don't break our hearts." The next morning, 10,000 people descended on the Capitol to speak at a public hearing on the bill. Every two minutes, a new speaker would tell legislators how the bill would have a devastating impact on them and those they care about. Since the bill couldn't go to a vote while the hearing was open, we made an urgent call to the community to come testify. Armed with their sleeping bags and pillows, people settled inside the Capitol building and waited for their opportunity to voice concerns about the proposed legislation. And so it began . . . we occupied the Wisconsin State Capitol.

As the Wisconsin Uprising gathered strength, hundreds slept on the marble floors at night, thousands packed the building during the day, and weekend rallies reached crowds of 200,000. Life as usual ceased; the Capitol became the center of our universe. Support flooded in—the Green Bay Packers (themselves union members) spoke out against the attack on workers, allies from 64 countries and all fifty states ordered pizza to feed protestors. People donated to the local grocery co-op for healthy food deliveries and locals arrived with all sorts of supplies. Actions became more creative everyday; signs had slogans like "Walker is a Koch head," a Star Wars AT-AT costume read "Imperial Walker," a Facebook invite by undergrads urged students to "sleep with your TA" at the Capitol.

The occupation wasn't planned, but it wasn't entirely spontaneous either. The TAA, and sociology students in particular, were informed on the issues and were connected to various community organizations. When the moment came, we tapped into our diverse resources and networks to facilitate action. We also had an amazing TAA steward (representative) for our department, Adrienne Pagac, who for years urged sociology grads to participate in the union and kept us informed of political events. She helped inspire a sense of urgency and convey the magnitude of the moment before us.

As sociology graduate students, we used our sociological training to carry out our work, adapting it to serve us in this moment of crisis. We contributed to messaging during the movement, contextualizing the struggle within labor history and social movements. The TAA established a control room in the capitol, which served as the launching pad for the occupation-inspired Web site, www.defendwisconsin.org. Members of the TAA coordinated food donations and project teams for everything from teach-ins to crowd marshaling to trash pickup. Many of us were teaching assistants (TAs), so we encouraged our students to witness democracy in action. At the time, I was a TA for sociology of race and ethnicity course and my students were learning about the civil rights movement. While they were reading about nonviolent civil disobedience in their textbook, many of their peers and TAs were deploying those same tactics just down the street in the occupation of their state capitol. Some of my own students brought their books to the Capitol and joined the Uprising.

Recognizing the role of coalitions in social change, I served as liaison between the TAA and various community groups who were an essential part of the occupation. My training in sociology, and in community-based research particularly, has taught me to pay attention to who has a seat at the table and who doesn't. Therefore, I pushed the groups I worked with to be more inclusive, to bring other (especially marginalized) groups into the decision-making process. The coordination between groups in the space of the Capitol was crucial, where everyone from big unions to media to concerned individuals worked toward a common goal: to maintain the occupation to defeat the bill.

Inside the capitol building, we had to share the place with police—lots of them. There was an odd solidarity between police and protestors for much of the occupation, as the Capitol and Madison police publicly refused to remove occupiers (they too were unionized employees after all). We experienced nothing like we all saw in the videos of police brutality against the Occupy Movement only months later. As hundreds of people continued living inside the Capitol, however, there was pressure by the Department of Administration to kick us out. Yet, there was constant negotiation over that space and creative attempts to resist eviction. When police expressed concern that the marble in the Capitol might be stained, we kept the red juices out and replaced the poster tape with removable blue painters' tape. We recruited legislative aides (who maintained access to the building when it was restricted for the public) to carry in food and people snuck it through windows. When police told us to stop opening windows, we told them to open another door. And so it went—this negotiation with the powers that be to ensure people were fed and healthy and able to respectfully practice their civil right of assembly. It was after all *our house* and we wanted to take care of it. The entire experience was a great exercise in the pragmatist approach, applying my sociological understanding of power to serve tangible goals . . . combining theory and practice.

These are the moments where real, long-lasting change takes root, where true shifts in power and consciousness begin. Change is unlikely to be immediate, but every person who saw that tremendous Capitol packed with everyone from children to grandmothers, from ironworkers to graduate students, and from librarians to cafeteria cooks, witnessed the power of collective action. The experience of thousands of Americans together, demanding and exercising their democratic rights, forever transformed all those involved in the Wisconsin Uprising. The police never did forcibly remove us. We ended the occupation after a court order ruled the passing of Act 10 unconstitutional. Of course, that ruling was later overturned by a divided Wisconsin supreme court. However, since that time, a federal court ruling made some elements of Act 10 unconstitutional again and a local judge overturned many elements of Act 10 for city, county, and school district employees. So the occupation ended, but the battle continues inside the courtroom and the Capitol, and on the ground.

Here in Madison, we never went back to life as usual—protestors maintain a daily presence at the Capitol. We continue to build community-based coalitions, many rank-and-file union members are organizing grassroots style, and the TAA itself has become more engaged with the broader community off campus. Many of the students in the UW-Madison sociology department remain heavily involved in the movement. The experience inspired some to work on public policy initiatives that will serve our goals, since a major lesson born out of Governor Walker's attack is the power and impact of public policy. What legislators do in those fancy rooms across this country, while seemingly boring, is what creates the reality we have to work within. If we want to change that reality, we need to create policies we believe in and hold legislators accountable to the people they serve.

As I write this, I still don't have my collective bargaining rights, healthcare is accessible to fewer people, and preK–12, technical college, and university funding has been slashed . . . we have lost many battles. But during those freezing days of February, Wisconsin started fighting back against decades of right-wing attacks on working people. Students helped spark the new social consciousness that is spreading across this nation and taking root in the Occupy Movement and a revived labor movement. We proved that a bunch of sociologists in action could transform the U.S. struggle for economic and social justice.

DEMOCRACY MATTERS: GIVING STUDENTS A POLITICAL VOICE

Joan D. Mandle

Democracy Matters, Hamilton, New York

Joan Mandle is the executive director of the national nonpartisan student organization Democracy Matters and professor emerita of sociology at Colgate University. She directed the college's Women's Studies Program from 1991 to 2000 and founded the Colgate Center for Women's Studies. Her teaching and extensive published work exploring social change and social movements include *Women and Social Change in America* (2nd ed., 1981) and *Can We Wear Our Pearls and Still Be Feminists: Memoirs of a Campus Struggle* (2000). She has taught sociology in Japan and China, and has been a Distinguished Visitor at both the University of California, Berkeley, and Mills College. In 1984 and again in 1986, Professor Mandle was campaign manager for incumbent congressman Bob Edgar of Pennsylvania.

Since the publication of C. Wright Mills's brilliant *Sociological Imagination* in 1959, many sociologists have explored the intersection of *private troubles* (biography) and *public issues* (social structure) (pp. 7–11). My own passion as a sociologist has been similarly focused—on understanding how people attempt to remedy their personal (private) concerns by challenging and changing society's institutions (social structures) through social movements. But following Marx's 1845 dictum not to just understand but also to change the world (1998), I have actively participated in movements for social change since 1964 when, as an undergraduate, I became part of the civil rights movement's Freedom Summer.

In the movements that followed—the antiwar, student, and antipoverty movements of the 1960s, and later the women's and environmental movements—students connected their private concerns about the draft, discrimination, or the destruction of the natural world with political efforts to change laws. My research followed these movements, tracking how students' personal concerns became transformed into political action. And in my own movement activism over the years, I have also personally worked toward the advancement of that political transformation.

Then, in 2000, I started a new venture with my adopted son, Adonal Foyle. While studying the civil rights and antiwar movements in his college sociology classes, Adonal had become convinced—as I was—that young people are the key to long-term social change. But at the same time, he was bothered that his friends' potential power for change was dissipated by the divisions among campus groups, each of which espoused its own separate cause. He was also concerned that so many students were cynical about and avoided political activism.

When Adonal was chosen as the eighth pick in the 1997 NBA draft, we suddenly had the resources to try to change that. Adonal decided to contribute part of his NBA salary to creating a brand-new, nonpartisan student organization, Democracy Matters (DM). Our goal was explicitly political. As Adonal put it, "We wanted to harness the collective power of young people, and make their voices heard politically so that they could become effective social change advocates for themselves and for others."

As the executive director, I ran the day-to-day operations of the organization, drawing on both my sociological imagination and what I had learned while studying social movements and social change. We offered Democracy Matters internships on college campuses, setting up a competitive process to award them to politically passionate and involved undergraduates. DM interns were charged with (1) creating DM chapters specifically dedicated to broad political discussion and engagement, (2) running prominent campaigns on campus to educate and involve others in political action for institutional change, and (3) creating on-campus coalitions as well as links between students and broader political organizations in their communities and the nation.

Through Democracy Matters, over 500 campuses and thousands of students have been involved in a social movement focused on changing our institution of government. DM organizers call for a more just and equal political system, pointing to the harmful role of special interests' big campaign contributions in our democracy. The goal of their activism and organizing is to create a public financing option for candidates for public office. This goal allows students to work for a specific political change while at the same time addressing the many causes and issues they care about—the environment, health care, tuition increases, foreign wars, and other hot-button issues, all of which are impacted by the vast political power of oil, banking, defense, and health care corporations as well as other wealthy campaign donors.

DM chapters have organized a wide variety of panel and brown-bag discussions, debates, and creative events on campus to help build the political power to win this change. They show how unfair the system of private campaign funding is that allows wealthy individuals to accrue disproportionate political power by funneling millions of dollars into political campaigns. In addition, they emphasize that those who can't make big donations (which is over 99% of Americans, and includes most students) have little say in their own government—the institution whose laws and policies affect them every day.

With the ever-growing costs of election campaigns (the 2008 elections were estimated to cost over $5 billion), politicians have become increasingly dependent on rich contributors, especially those representing wealthy corporations and big special interests. These donors gain enormous influence over who runs, who wins, what issues make it to the political agenda, and how politicians vote once elected. Unfortunately, young people have become increasingly cynical about government because they, like most Americans, feel they lack a political voice and could never afford to run for office. They believe that politicians are bought and sold, and that elections are auctions going to the highest bidder. Democracy Matters chapters offer students a chance to make a difference—to join with others in political action to deepen democracy by, as our slogan states, "Getting big private money out of politics and people back in."

The issue of money and politics attracts many student activists because it connects their personal concerns with an understanding of institutional processes. In their organizing and educational efforts, DM members link the causes that students care to the distorting role of big private money in politics. For example, some chapters have created educational poster campaigns, class raps, and newsletters pointing out how environmental policy is influenced by corporate donations from oil and energy companies, or how health care reform has been historically opposed and more recently shaped by huge contributions from the health care and pharmaceutical industries. By writing articles for their school paper or screening relevant films, others have made their campus communities aware that the availability and cost of student loans are affected by campaign donations from big banks and lending companies, or that money donated to campaigns by the defense industry affects the character of our foreign aid and foreign policies around the world.

Democracy Matters students implement other carefully designed grass-roots campus campaigns as well, in order to build coalitions and a broad movement for change. At Ohio State, for example, DMers organized an outdoor carnival, inviting other campus groups to join in creating booths and games that not only were fun but also raised awareness by offering information on issues and causes. They had several student bands entertain, calling the carnival "Amps Against Apathy," and gave each organization a few minutes to talk about its issues. As the host organization, Democracy Matters then had a chance to explain how all these issues were affected by money in politics, and how together they could fight political apathy. At Colgate, DMers collaborated with campus environmental groups to write a skit that they performed on the quad. In front of a giant cutout of the Capitol with a "For Sale" sign across it, students holding logos of big oil and energy corporations stuffed a smiling "Fat Cat" politician's pockets with fake money tied to strings, while the Fat Cat turned his back on other actors with signs depicting "wildlife refuge," "student," or "environmental advocate."

In addition to increasing the understanding of how private funding of election campaigns harms both our democracy and the issues students care about, Democracy Matters activists engage students in actions to promote public financing as an alternative. Today in a number of states like Connecticut, Maine, and Arizona, laws already allow political candidates to choose to run without raising big-dollar contributions. These laws create voluntary public financing systems that offer ordinary citizens equal resources to run their campaigns without being indebted to big funders and special interests. Legislation pending in Congress and in many states would create similar institutional change.

Though these laws are fiercely opposed by wealthy special interests, Democracy Matters activists—along with our allies from other national organizations—have fought back. Students at DM campuses join petition and letter-writing campaigns, urging elected officials to support public financing. Others send faxes, make calls to politicians' offices, or design and mail homemade cards—for example, Halloween cards with jingles about how "scary" big-money donations are. DMers often personally deliver these signatures, cards, or letters at meetings with their congressional representatives to discuss the need for a fairer system of campaign financing.

Democracy Matters is just one of many organizations that uses—some without realizing it—the sociological imagination to motivate and activate efforts to change political institutions. One exciting outcome is that the students themselves are often changed in the process. I want to close by sharing some quotes from Democracy Matters activists, and by urging you to use your own sociological imagination to join with others in making our political institutions more just.

Democracy Matters has been a huge light on my path of activism for social justice. I got involved as an advocate for public funding of campaigns to try to heal the ills of our society. I can see no greater cure for the disease of war, poverty, and greed. —A. W., LA Valley College (CA)

Democracy Matters taught me how to truly be an effective political activist and make a difference. —Z. L., Winston Salem University (NC)

Democracy Matters made me understand that my worries about paying for college and getting a decent-paying job and health insurance after I graduate next year were part of a bigger political picture tied to the need for social change. —A. C., University of Minnesota

References

Marx, K., & Engels, F. (1998). *The German ideology, including theses on Feuerbach (Great Books in Philosophy)*. New York, NY: Prometheus. (Original work published 1845)

Mills, C. W. (1959). *The sociological imagination*. Oxford, UK: Oxford University Press.

DISCUSSION QUESTIONS

1. Do you agree with Alon, that institutions of higher education should establish class-based affirmative action programs? Why or why not?

2. Think back to how you prepared for the college admissions process. In what ways did your social class background impact (a) how/if you studied for standardized tests, (b) to what schools you applied, and (c) where you decided to attend college?

3. Dadit Hidayat, Randy Stoeker, and Heather Gates describe the collaborative efforts of a university and a community organization to promote water sustainability in their region. Have you ever thought about how your college or university could support the work of activist groups in the community and help to create social change? What might be the benefits of such collaborative efforts for (a) the community groups, (b) the college or university, and (c) the students involved? How do you think you could use your sociology skills to help your college or university to determine the community groups with which it should partner? Why?

4. Why do Hidayat, Stoeker, and Gates think that project-based research is the best means for colleges and universities to support community action? Does their argument make sense to you? Why or why not?

5. Thanks to their sociological training, John O'Brien and Besheer Mohamed were able to work to counter the stereotypes that rose up about Muslims after September 11, 2001. How did they do so? What stereotype, if any, would you most like to counter? Why?

6. Based on what you read in O'Brien and Mohamed's piece, describe how religion can (a) unite people and (b) divide people. In your opinion, is religion more of a unifying or dividing force in the United States today? Why? Be sure to thoroughly explain and support your answer.

7. Charity Schmidt describes how, during the Wisconsin Uprising, she and fellow sociology graduate students "used [their] sociological training to carry out our work, adapting it to serve us in this moment of crisis." Describe two ways Charity used sociological tools in her efforts to support the Wisconsin Uprising. Do you think their sociological training was particularly useful in these efforts? Why or why not?

8. Imagine that you were in a situation similar to the one Schmidt found herself during the time of the Wisconsin Uprising. Your governor has just threatened to pass a bill that you believe will cause serious harm to millions of people in your state (yourself included). How would you react? Why? What sociological tools *could* you use to influence your elected leaders to stop or mitigate the harm of the proposed bill? *Would* you use these tools? Why or why not?

9. When Adonal Foyle went to college, he was disheartened by the political apathy of many of his fellow students. Are you politically active? Why or why not? Do you think most students on your campus are politically active? If yes, what do you think motivates their activism? If no, how might you explain their lack of political activism?

10. Mandle and her son created Democracy Matters for the purpose of encouraging more student engagement around political issues. If your college president asked you to use your sociological eye to figure out how politically active the student body at your campus is and how to make it more active, how would you go about carrying out this assignment?

11. Do you consider yourself to be an educated citizen who knows enough about how your society works to influence it? Why or why not? Has your college education (thus far) helped prepare you to become a knowledgeable, effective citizen? If so, how? If not, why not?

12. How might you describe the relationships among higher education, government, and religion? Do you think any of these social institutions can work without the other two? Why or why not?

RESOURCES

The following Web sites will help you to further explore the topics discussed in this chapter:

The American Democracy Project	http://www.aascu.org/programs/adp/
ASA Section on Sociology of Education	http://www2.asanet.org/soe/
ASA Section on Sociology of Religion	http://www2.asanet.org/section34/
Association of Religion Data Archives	http://www.thearda.com/
CIRCLE: The Center for Information & Research on Civic Learning and Engagement	http://www.civicyouth.org/
Education Resources Information Center	http://www.eric.ed.gov/
International Bureau of Education	http://www.ibe.unesco.org/en/services/
Sociosite Power: Conflict and War	http://sociosite.net/topics/power.php
ASA Section on Political Sociology	http://www2.asanet.org/sectionpolitic/
Sociological Tour Through Cyberspace Political Science	http://www.trinity.edu/~mkearl/polisci.html
Sociosite Religion and Spirituality	http://sociosite.net/topics/religion.php
Virtual Religion Index	http://virtualreligion.net/vri/

To find more resources on the topics covered in this chapter, please go to the Sociologists in Action Web site at **www.sagepub.com/korgensia2e**.

Index

About the Editors

Kathleen Odell Korgen, PhD, is Professor of Sociology at William Paterson University in Wayne, NJ. Her primary areas of specialization are race relations, racial identity, and public sociology. Professor Korgen's publications include *The Engaged Sociologist: Connecting the Classroom to the Community* (with Jonathan White) (Pine Forge 2011), *Multiracial Americans and Social Class* (Routledge 2010), *Crossing the Racial Divide: Close Friendships Between Black and White Americans* (Praeger 2002), and *From Black to Biracial* (Praeger 1999).

Jonathan M. White, PhD, is Associate Professor of Sociology at Bentley University and Director of the Bentley University Service Learning Center. He specializes in inequality, globalization, human rights, and public sociology. He has received numerous teaching and humanitarian awards, is founder of Sports for Hunger, and has served on the board of directors of Free the Children, the Graduation Pledge Alliance, Me to We, Youth for Peace, and other civic engagement organizations. He is co-author of *The Engaged Sociologist: Connecting the Classroom to the Community* (with Kathleen Odell Korgen) (Pine Forge 2011) and served as associate editor to *The New York Times* bestseller *Me To We* by Marc and Craig Kielburger. Dr. White lives in Massachusetts and is the proud uncle of 13 nieces and nephews.

Shelley K. White, PhD, MPH, is Assistant Professor of Public Health at Worcester State University. She recently served as Instructor of Sociology and Public Health at Simmons College, where she also coordinated the Simmons World Challenge, an interdisciplinary, student-led learning program which creates actionable solutions to global social justice issues. Shelley's teaching and research focus on health and illness, globalization and development, inequalities, social movements and public sociology.

She previously worked in HIV/AIDS policy globally and domestically, and serves on the board of directors of Free the Children and SocMed. Her recent publications appear in the *Journal of Human Rights Practice, Education, Citizenship and Social Justice,* and *Critical Public Health.*

ⓢSAGE research methods

The essential online tool for researchers from the world's leading methods publisher

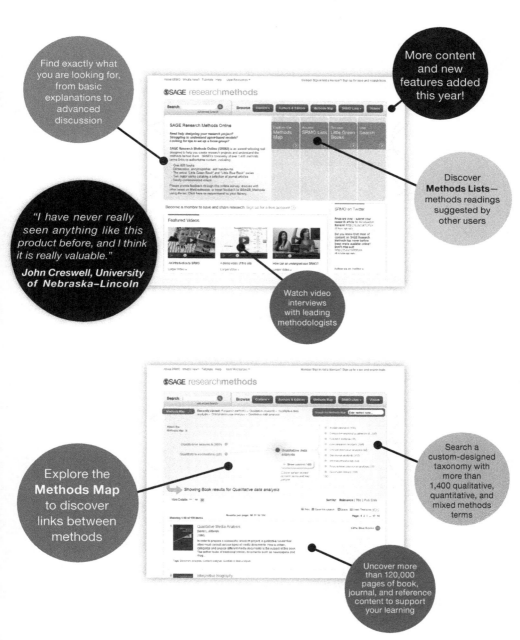

Find exactly what you are looking for, from basic explanations to advanced discussion

More content and new features added this year!

"I have never really seen anything like this product before, and I think it is really valuable."

John Creswell, University of Nebraska–Lincoln

Discover **Methods Lists**— methods readings suggested by other users

Watch video interviews with leading methodologists

Explore the **Methods Map** to discover links between methods

Search a custom-designed taxonomy with more than 1,400 qualitative, quantitative, and mixed methods terms

Uncover more than 120,000 pages of book, journal, and reference content to support your learning

Find out more at
www.sageresearchmethods.com